INVESTIGATING HUMAN INTERACTION THROUGH MATHEMATICAL ANALYSIS

Investigating Human Interaction through Mathematical Analysis offers a new and unique approach to social intragroup interaction by using mathematics and psychophysics to create a mathematical model based on social psychological theories.

It draws on the work of Dr. Stanley Milgram, Dr. Bibb Latane, and Dr. Bernd Schmitt to develop an algebraic expression and applies it to quantitatively model and explain various independent social psychology experiments taken from refereed journals involving basic social systems with underlying queue-like structures. It is then argued that the social queue as a resource system, containing common-pool resources, meets the eight design principles necessary to support stability within the queue. Making this link provides a means to advance to more complex social systems. It is envisioned that if basic social systems as presented can be modeled, then, with further development, more complex social systems may eventually be modeled for the purpose of identifying and validating social structures that might eventually support stable governments in our common environment called Earth.

This is fascinating reading for academics and advanced students interested in political theory, detection theory, social psychology, organizational behavior, psychophysics, and applied mathematics in the social and information sciences.

Kurt T. Brintzenhofe is a retired Navy officer, civilian systems analyst, and system-of-systems major acquisition test engineer/director who, with a master's degree in operations research, worked for many years in mathematical modeling and operational evaluation. He is currently continuing his education as an advanced graduate student at the University of Maryland, where his research interests cover applied mathematics, history, social psychology, and psychophysics.

INVESTIGATING HUMAN INTERACTION THROUGH MATHEMATICAL ANALYSIS

The Queue Transform

Kurt T. Brintzenhofe

Routledge
Taylor & Francis Group

NEW YORK AND LONDON

Designed cover image: Getty

First published 2023
by Routledge
605 Third Avenue, New York, NY 10158

and by Routledge
4 Park Square, Milton Park, Abingdon, Oxon, OX14 4RN

Routledge is an imprint of the Taylor & Francis Group, an informa business

© 2023 Kurt T. Brintzenhofe

Library of Congress Cataloging-in-Publication Data
Names: Brintzenhofe, Kurt T., author.
Title: Investigating human interaction through mathematical analysis : the queue transform / Kurt T. Brintzenhofe.
Description: New York, NY : Routledge, 2023. | Includes bibliographical references and index.
Identifiers: LCCN 2022027832 (print) | LCCN 2022027833 (ebook) |
    ISBN 9781032350745 (hardback) | ISBN 9781032350714 (paperback) |
    ISBN 9781003325161 (ebook)
Subjects: LCSH: Social interaction—Mathematical models. | Queuing theory. |
    Abelian groups. | Algebra, Abstract. | Psychophysics. | AMS: Mathematical logic and foundations. | General algebraic systems. | Measure and integration.
Classification: LCC HM11111 .B755 2023 (print) | LCC HM11111 (ebook) |
    DDC 302.01/5118—dc23/eng/20220822
LC record available at https://lccn.loc.gov/2022027832
LC ebook record available at https://lccn.loc.gov/2022027833

ISBN: 978-1-032-35074-5 (hbk)
ISBN: 978-1-032-35071-4 (pbk)
ISBN: 978-1-003-32516-1 (ebk)

DOI: 10.4324/9781003325161

*In memory of Ladd Zastoupil, a high school sociology/
psychology teacher who made a difference.*

CONTENTS

FIGURES

PREFACE

When first starting out, it was thought this effort would possibly result in a journal article, but it grew as underlying social patterns were observed, and subsequent equations supporting further investigation were derived. The decision to provide the book as open access under copyright was made for two reasons. The first being that getting time from the academic community for review and interaction has proven extremely difficult, so being open access may improve the opportunity for academic interaction from students and possibly professors. Second, being open access allows for anyone to have access, particularly students from diverse backgrounds who have the energy and sense of adventure to consider new ideas. In return, I ask and encourage those reading this book to provide constructive comments relating to necessary corrections or to provide ideas for further development and potential collaboration. The objective is improvement, not laurels or stones.

This book provides the basis for a repeatable and testable quantitative approach to the analysis of hierarchal social systems. To that end, beginning with the derivation of the modified Weber's Law and Fechner's Law, an algebraic structure is carefully developed ultimately leading to the definition of a social algebraic Abelian group in Chapter 6. Defining an Abelian group is not the end of this effort; it is instead the foundation on which to begin. Whether all of the equations and theorems throughout this book stand the test of time, at least they are testable, and more importantly, they may be corrected when necessary and further developed as we learn more. As an analyst with 30 years of applied operational experience, there are no illusions that improvements and corrections to this work will not be identified. Irrespective of professional embarrassments that may arise from reader reviews, experience dictates that anything of potential worth must eventually be placed at the mercy of the reader if improvement is to occur. It is the diverse view and expertise of the reader that can make the most significant contributions, and, unlike the author, the reader has not invested 4 years of his or her life on the subject which can lead to an inherent bias or blindness toward certain areas of this book. Please consider all of these factors as you read.

The reason for devoting 4 years to reading and developing the ideas contained within this title is simple. History continues to repeat itself, and if we are to progress as a species toward something ultimately resembling a civilized society, we need to understand why this repetition occurs and how to break the seemingly continual cycle of war, famine, and social displacement. Technology alone will not solve these problems. Our species is a primitive species with a significant amount of social evolution necessary before we may claim the title of being civilized, and, in the meantime, we need to find ways to increase the likelihood of our survival as a global community until we are able to sustain peaceful coexistence without conscious effort.

ACKNOWLEDGMENTS

This book has its genesis from my 1977 high school sociology class, taught by Mr. Ladd Zastoupil. His introduction on the work of Dr. Pitirim Sorokin changed how I viewed the world and the people in it. To my graduate social psychology professor Dr. Charles Stangor, who introduced me to Leon Festinger's work in 2017, this significantly updates my final report.

With that, to the teachers of the world, I say thank you. Education enables us to observe events around us more broadly and efficiently, formulate opinions based on the past work and experience of others, and focus our questions as a result to better understand the world as it really is, not as a few may desire to portray it for their own benefit.

I would also like to thank Ms. Michelle Marques, daughter of Dr. Stanley Milgram and personal representative of the Alexandra Milgram Estate which owns the copyright to the Stanley Milgram Papers held at the Manuscript and Archives, Yale University Library. Ms. Marques has been generous in her responses and allowed the use of raw data from the Stanley Milgram Papers for this book. The analysis to be presented would not have been possible without the data. It is a true gift to obtain the necessary level of information typically absent in journal articles.

For permissions that enliven and make this book possible, I would like to thank:

- Stanley Milgram Papers (MS 1406), Manuscripts and Archives, Yale University Library, with copyright permission from the Estate of Alexandra Milgram.
- Taylor & Francis STM partners including the American Psychological Association, Cambridge University Press, Elsevier Publishing, McGraw-Hill Professional, and SAGE Journals.
- Digireads.com for responding and confirming their material was public domain.
- United States Navy Office of Naval Research for its funding of early research and development at universities across the country.
- Stanford University Press for keeping Leon Festinger's work alive and for allowing the necessary copyright permission.

INTRODUCTION

This is a book written for advanced undergraduate and graduate university students having an interest in crossing the academic boundaries of social psychology, psychophysics, political economics, information theory, signal detection theory, and various disciplines of mathematics along with a desire to constructively consider alternative social structures, such as Elinor Ostrom (1990) proposed and defined within closed social systems relying on limited resources. Earth is a closed social system relying on limited resources. It is a large ball, with water and a thin currently breathable atmosphere, surrounded on all sides by the vacuum of space. Given the growing political polarization of nations, the dynamic demographics due to disasters (pandemics included) and conflicts, and the emerging climate crisis which will amplify those disasters and conflicts, it would seem appropriate to systematically consider both the continued stability of existing social structures and alternative nonnormative social structures. This book only considers western social systems (Mann, 1969, p. 349) and their structures as a basic algebraic social subgroup (Herstein, 1999, pp. 41 & 51). Incorporating social systems of other cultures is critical, but well beyond the capability of this author, existing data, and this initial effort.

Once past this introduction, a brief but necessary foundation based on psychophysics and cognitive dissonance theory (Festinger, 1957) is provided. Weber's and Fechner's Laws are constructively modified to reflect empirical data near absolute sensory threshold. With these provided as the mathematical foundation, the model evolves further in Chapters 2 and 3, maturing in Chapter 3 using field experiment data involving queues. Data sets from past laboratory and field experiments having an underlying queue-like structure are then modeled in Chapter 4 to compare empirical probability of reaction with theoretical probability of reaction as a function of sensation magnitude (Festinger, 1957, p. 181) for a given stimulus and social situation. The result is nothing more than an operational form of Kurt Lewin's general behavior model concept (Burnes and Cooke, 2013, pp. 412–413) where:

Dissonance sensation magnitude = f (average person, social situation, stimulus)

DOI: 10.4324/9781003325161-1

for a to-be-defined function f. The model developed here, thereby defining such a function f, is unlikely to be the end to mathematical development in this area; in fact, it would be disappointing if it were. Instead, it is considered a bold alternative approach to years of qualitative theories that may or may not build upon each other toward a common goal (Pettigrew, 1986) or which can be tested in a manner supporting accurate evaluation of cause and effect (Funder et al., 2013). More importantly, it proposes a basic algebraic subgroup – that I shall enjoy labeling a social subgroup – which, if the model is sufficiently general, may be improved upon and expanded to better represent not only western social systems but eventually all interacting social systems as well. In effect, it offers a new way to consider basic interactive group dynamics based on Fechnerian psychophysics and cognitive dissonance theory while addressing and not excluding Stevens' power law (Stevens, 1957).

With so much time put into this effort, the recurring question in my mind has been whether it will be useful or not. The answer to this question is if it may be used to make the world a better place through a better understanding of social interaction, then it is useful. If it turns out to not be useful to the audience, even then I will have learned an extraordinary amount and am personally satisfied. But, if it is abused by those who seek power and status for their own benefit, then should this be published? This would seem to be the dilemma that confronted Kurt Lewin in his 1939 article (Lewin, 1939/1997):

> It is a commonplace of today to blame the deplorable world situation on the discrepancy between the great ability of man to rule physical matter and his inability to handle social forces. This discrepancy in turn is said to be due to the fact that the development of the natural sciences has by far superseded the development of the social sciences.
>
> No doubt this difference exists, and it has been and is of great practical significance. Nevertheless, I feel this commonplace to be only half true, and it might be worthwhile to point to the other half of the story. Let us assume that it would be possible suddenly to raise the level of the social sciences to that of the natural sciences. Unfortunately, this would hardly suffice to make the world a safe and friendly place to live in. Because the findings of the physical and the social sciences alike can be used by the gangster as well as by the physician, for war as well as for peace, for one political system as well as for another. (loc. 1469, with permission from the American Psychological Association)

The self-serving politician, many of the elite in every society, or the gangster's use of the social sciences has and still occurs. Though the empirically based results in this book are stimulating, they merely give a mathematical form to what is already being put into daily use by politicians, our social and economic elite who influence the information we receive, and organizational behaviorists who, like the alchemists of old, work for the social and economically elite to increase their wealth. By giving quantitative form to existing behavior, what has been done in relative darkness is now, partly at least, brought out more clearly into the light of day. Furthermore, given the condition of this world as it exists today, both socially and environmentally, there is critical need in the near term for new tools and thoughts to create, consider, and systematically investigate informed alternative approaches which may better direct our social energies toward a sustainable future. That is the motivation and intended objective of this book.

With that in mind, when reading the contents contained here, do not read the equations as a means to solve some exercise problem in a textbook. Instead, view them as a means to frame and dissect the basic social interactions taking place and consider how they may further develop to allow for the analysis of more complex social systems. It is difficult to know where to go from here, but selecting an interesting location in social space and just moving forward as proposed have their advantages for the adventurous.

References

Burnes, B., & Cooke, B. (2013). Kurt Lewin's Field Theory: A Review and Re-Evaluation. *International Journal of Management Reviews, 15*, 408–425. https://doi.org/10.1111/j.1468-2370.2012.00348.x

Festinger, L. (1957). *A Theory of Cognitive Dissonance*. Stanford University Press. Retrieved from www.sup.org/books/title/?id=3850

Funder, D., Levine, J., Mackie, D., Morf, C., Sansone, C., Vazire, S., & West, S. (2013). Improving the Dependability of Research in Personality and Social Psychology: Recommendations for Research and Educational Practice. *Personality and Social Psychology Review, 18*(1), 3–12. https://doi.org/10.1177/1088868313507536

Herstein, I.N. (1999). *Abstract Algebra*. John Wiley & Sons.

Lewin, K. (1939/1997). Experiments in Social Space. In G.W. Lewin (Ed.), *Resolving Social Conflicts and Field Theory in Social Science*. American Psychological Association (Kindle Edition). Retrieved from www.apa.org/pubs/books/4318600

Mann, L. (1969). The Waiting Line as a Social System. *American Journal of Sociology, 75*(3), 340–354. https://doi.org/10.1086/224787

Ostrom, E. (1990). *Governing the Commons*. Cambridge University Press. https://doi.org/10.1017/CBO9781316423936

Pettigrew, T.F. (1986). The Intergroup Contact Hypothesis Reconsidered. In M. Hewstone & R. Brown (Eds.), *Contact & Conflict in Intergroup Encounters* (pp. 169–195). Basil Blackwell (John Wiley & Sons).

Stevens, S.S. (1957). On the Psychophysical Law. *The Psychophysical Review, 64*(3), 153–181. https://doi.org/10.1037/h0046162

1

SOCIAL PSYCHOLOGY AND PSYCHOPHYSICS

Laying the Foundation

Thomas Pettigrew, protégé of Gordon Allport, wrote a compelling commentary in 1986 regarding the scope of investigation being pursued by theoretical social psychologists (Pettigrew, 1986, pp. 169–171). Allusion was made to the multitude of theories in social psychology such as cognitive dissonance theory, attribution theory, social impact theory, contact theory, social identity theory, and the list goes on. A few of the many questions resulting from his expose are whether social psychological theories are exclusive to one another, whether they correct, and, if combined, do they support a greater understanding of how we interact within a social system? If they are correct and do support a greater understanding, then how do we prove it across the many situations and cultures? These thoughts identify the need to reconsider how both theory and models are being used – a reconsideration that leads to addressing the difficulty of proving and combining much of the existing theoretical work within social psychology – one which encompasses the greater context of various existing social situations but directly addresses the need and means to go beyond theories and models which may work within western culture but may not be representative of other equally important cultures. Though it seems progress has been made since 1986 when Thomas Pettigrew wrote his expose, the field of social psychology still seems to lack a common testable integrated structure on which to build upon and support major cross-cultural advances in the field. The goal of this book is to propose an approach to meet the need, a quantitatively testable and methodical approach which may offer a means toward the investigation of increasingly complex social systems in both western and ultimately non-western cultures.

What is proposed as a starting point, harkening back to Kurt Lewin and his Field Theory while relating results to psychophysics as presented by Gustav Fechner and Stanley Stevens, is a systematic mathematical approach to quantitatively model social interaction within social groups operating as basic social systems. The proposed model developed here addresses behavior in the form of mean probability of reaction as a function of the specific social situation and given social event. It is based on group member social position, while currently constrained to associated western

DOI: 10.4324/9781003325161-2

social norms due to lack of data from other cultures. The approach is derived initially from quantitative data resulting from past independent field experiments conducted in New York City and documented in Milgram et al. (1969), Milgram et al. (1986), Schmitt et al. (1992) and supplemented by data from the Stanley Milgram Papers (MS 1406), Manuscripts and Archives, Yale University Library. What is to be the basis for this proposed social model is one of the most fundamental of western embryonic social systems (Mann, 1969) – the first come, first served queue.

As in abstract algebra, where a proper subgroup possesses some or all of the mathematical operations of the group it operates within (Herstein, 1999, p. 51), if a basic social system existing within a larger more complex social system is mathematically representable, then the potential exists to use that basic mathematical representation to better understand the more complex social system surrounding it. This in turn could lead to further mathematical development and afford the opportunity for various existing social psychological theories to identify with common parts of a greater theoretical whole, thus allowing for more significant advances by systematically weaving various related research efforts into a larger tapestry, creating a more comprehensive and dynamic scene instead of being restricted to the view of many seemingly unrelated parts (Becker, 1934, p. 399).

1.1 Proposing an Objective for Social Psychology

As in trying to define sociology (Lenski, 2005, p. xiii), there may be some difficulty in trying to accurately define social psychology and why it is important. In defining social psychology, Charles Stangor noted at the beginning of his teaching career the challenge of trying to convey to his students a coherent structure in which to place the many topics of social psychology (Stangor et al., 2014, p. ix). A strict definition of social psychology will not be pursued here since that would extend beyond the scope of this effort and likely create a needless distraction. Fortunately, most definitions for social psychology revolve around how we react to and are influenced through social interaction with others, and that is enough if only one extra detail is added. As Kurt Lewin noted (Lewin, 1935/1997, p. 107), to understand how others affect us, it is necessary to include the background or social situation in which the interactions of interest are occurring. Even with the inclusion of social situation into the definition, a clear objective still seems to be lacking as to why social psychology is being pursued or how results obtained will be made useful to the general population who have and continue to expect tangible benefits from efforts by those of the academic community. To address this perceived shortfall and establish the context of this work, a single typed sentence on a piece of scratch paper attributed to Stanley Milgram is resurrected.

In the Stanley Milgram Papers (MS 1406), Manuscripts and Archives, Yale University Library, there exists a single sheet of paper having one typed sentence with corrections. The paper indicates, assuming Stanley Milgram typed it, that he may also have been considering the objective of the social sciences, and possibly social psychology in particular. On it is the sentence (Stanley Milgram Papers, MS 1406, Series No. 2, Box 24, Folder 8, with permission), *"Do the social sciences have anything of value to offer the world that can help preserve peace and prevent wars."* In that question is a core objective, one that I think every definition of social psychology should be including. If we continue to more accurately model social interaction in a quantitative, data-driven, and repeatable fashion, maybe the social, political, and economic sciences can better work together in improving approaches for conflict resolution and the identification of parameters supporting long-term stability after a conflict. One proposed supporting pillar of this objective is the work of economist Elinor Ostrom (Ostrom, 1990) who considered the development of theories for explaining and anticipating behavior under a multidisciplinary multi-tier framework as being one of the core challenges toward establishing sustainable social–ecological systems (Ostrom, 2012, pp. 69–70).

Every objective needs a strategy to get there. The strategy proposed here is to begin at the basic level by quantitatively defining and modeling social interaction within the fundamental social system of a western queue. The queue has an established set of social norms necessary for controlling conflict and a simple means of monitoring to maintain group integrity (Mann, 1969, p. 349). If the simple social system defined by Mann (1969, p. 349) for a given culture may be modeled, and that social system is representative of at least part of the culture and its social norms, then it is argued and demonstrated in Chapter 4 that the model may be applied to related group situations not involving a queue – but which have an underlying queue-like structure with similar social norms. Since it cannot be overemphasized, this is not to imply non-western queues or their associated social systems are unimportant or cannot be modeled, but that they have not been studied to the extent of western queues (Gillam et al., 2014). Non-western queues have dissimilar social norms reflecting the history and culture which they reside within (Gandhi, 2013), but these variations must be reflected in any large-scale theory that is eventually directed toward facilitating the ultimate goal of global stability. Bringing this all together, if we can achieve the necessary means to accurately monitor and, through model validation with increasing social complexity, establish rules of governance in which all can agree to in an effort to maintain group integrity (Ostrom, 1990), then stability may eventually be possible.

Before stepping through the basic quantitative model derivation for the queue as a social system, some background theories in social psychology must be provided and concepts introduced. Let us begin with a quick overview of relevant work and theories from Ernst Weber, Gustav Fechner, Kurt Lewin, Leon Festinger, and Stanley Stevens.

1.2 Cognitive-Dissonance-Based Definitions and Social Axioms

Field theory, influenced by Gestalt psychology, was developed by Kurt Lewin over a period of 25 years beginning in the 1920s. Briefly, his theory postulates the possibility to understand, predict, and identify the basis for changing individual and group behavior by identifying the psychological forces influencing an individual's or a group's behavior at a specific point in time. As it developed under Lewin, field theory shifted toward the analysis and modification of group behavior. It is argued by Burnes and Cooke (2013) that the decline of field theory after Lewin's death in 1947 was due mainly to his pursuit of mathematical rigor at the expense of practical relevance. This pursuit, it is further argued, tilted the balancing act necessary to maintain methodological soundness while remaining useful to the operational practitioner (Burnes and Cooke, 2013, p. 409). What follows in this book relies on a set of basic mathematical tools taken from calculus, probability theory, set theory, queueing theory, detection theory, and information theory. Effort will be placed on maintaining a balance between mathematical rigor and practical relevance for the general reader, but in the end, if something cannot be quantified, it cannot be measured, and if it cannot be measured, it cannot be tested and compared; so familiarity with some basic and advanced applied mathematical tools beyond the traditional and controversial hypothesis testing approaches is necessary (Wasserstein and Lazar, 2016).

An important equation by Lewin, providing a framework which with modification is presented here, is the person–situation interaction equation (Lewin, 1940/1997, p. 187; Burnes and Cooke, 2013, pp. 412–413),

$$Behavior = f[person, \ social \ situation, \ stimulus].$$

The original variable *environment* has been replaced with *social situation* to be clearer. Additionally, the variable *stimulus* is added since we will be interested in modeling probability of reaction to any relevant social stimulus interpreted as a deviation from accepted social norms or beliefs for the given social situation (Lewin, 1944/1997). If the person is in a group, then the behavior of other members in the group is also part of this behavior equation, but it is felt that combining it with *social situation* as done here reduces confusion until the exact meaning is clarified in Chapter 3.

As noted in Burnes and Cooke (2013), a person has multiple identities depending upon what group he or she is a member of at the time, and each group identity has its own social norms or beliefs. So, as Hogg and Terry (2000) argue, in what may be viewed as a more detailed and structured approach, people will react to relevant environmental stimuli as appropriate to the social norms of the group they are in at the time. If they do not, then they are considered deviant to some degree, resulting in penalties up to and including exclusion from the group (Hogg and Terry, 2000, p. 127). So, if we accept that group or individual behavior is a function of the group's social norms, social situation, and relevant social stimuli, the work of Leon Festinger may now be introduced.

Leon Festinger did his graduate work under Kurt Lewin at the University of Iowa, obtaining his PhD in 1942 on child behavior. One of his major works is the theory of cognitive dissonance (Festinger, 1957) which builds on the work of Kurt Lewin. Dissonance as used here involves an individual's emotional discomfort or tension brought about by any internally or externally initiated deviation from a group's social norms of which the affected individual currently considers his or herself a member. Three concepts introduced by Festinger are of significance here: 1) His two hypotheses of cognitive dissonance; 2) the concept that dissonance from a social stimulus is a function of its relevance to the situation and its extent deviation from cultural/social mores (see also Brauer and Chekroun, 2005); and 3) the implications of group identity and member interaction regarding the magnitude of dissonance felt and subsequent social pressures to reduce it. Making use of these concepts in a quantitative setting will require the introduction of a social space.

To operate within a mathematical framework requires a defined space. In Euclidian space, we have integers as a subgroup of real numbers which are a subgroup of imaginary numbers which are all part of complex space. To model social behavior, we will use the term social space in a manner consistent with that provided by Pitirim Sorokin (Sorokin, 1959, pp. 4 & 6) and Kurt Lewin (Lewin, 1939/1997, p. 58). To create the algebraic foundation for social space, we must provide adequate definitions and then state necessary axioms, where axioms are statements assumed as true (Smith et al., 2015, p. 28). Keeping in mind that any group member can be operating within multiple groups (subgroup of a larger group for instance) at any one time, the basic definitions to begin with are:

Definition 1. **Social Situation**: With slight modification, this is interchangeable with the term "background" as used by Lewin (1935/1997, p. 107). He indicates every action one performs within a specific social situation is determined by that situation. He goes on to clarify how interpretation may vary based on the social situation, where a statement or a gesture which may be quite appropriate between group members in a one social situation may be out of place or even insulting to group members in another. Our frame of reference depends on our culture and the social situation we are in, and that frame of reference dictates how we should react to relevant social stimuli that we are aware of at the time.

Definition 2. **Social Norms**: Shared beliefs, feelings, and reactions among group members perceived as appropriate for the given social situation. For a given situation, McDonald and Crandall (2015) indicate social norms provide an expectation by group members regarding what is appropriate behavior within the group context. Social norms should be consistent with associated cultural values.

Definition 3. **Social Space**: Group interaction of two or more people conforming to social norms associated with the situation(s) they are acting within at the time. Disturbances in social space result from group members experiencing one or more relevant social norm deviations for the given situation they are in.

Definition 4. **Social Relevance**: Any event, for a given social situation and time, which is within the realm of content matter to which an individual or group is affected by or concerned with to some level – up to and including direct and immediate impact. Social distance as used by Sorokin (1959, pp. 4–7) and Allport (1979, pp. 38–39) is a somewhat synonymous term.

Definition 5. **Extent Social Deviation**: The importance of unwanted consequences (Cooper, 2007, loc 2491) caused by a social deviation contradicting expected social norms or accepted beliefs of group members within a specific social situation.

Definition 6. **Social Dissonance**: An individual's emotional discomfort or tension brought about by experiencing, observing, or learning of an event exhibiting some amount of social deviation from accepted norms and having some level of relevance for the social situation in which the affected individual(s) involved are currently in.

Definition 7. **Consonance**: A pleasant or agreeable feeling brought about by lack of dissonant events for the given social situation.

Definition 8. **Cohesion**: This is a term discussed by Festinger (1957, p. 180) and may be described as the strength of social attraction felt between group members based on the cumulative importance of their shared social norms.

Definition 9. <u>**Social Noise**</u>: Any sensed information inhibiting the correct interpretation or reaction to the social event for the given social situation.

Definition 10. <u>**Reaction**</u>: Occurs when a group member, anticipating overall dissonance reduction as a result, makes the choice to perform a normatively appropriate (for the situation) dissonance reducing social action.

Sometimes dissonance cannot be immediately reduced when constrained by a restraining force (Lewin, 1997, pp. 101, 291, 316), and assuming cognitive adjustment to the situation is not acceptable. Once a choice does present itself in such a circumstance, where the existing dissonance may be reduced or removed at acceptable risk, then there will be a high probability of reaction to reduce the dissonance that is commensurate with the risk involved. A wonderful story, as communicated by the great Greek historian Plutarch, adds color to this concept, including Definitions 6 and 10:

> Chiomara, the wife of Ortiagon, was captured with the other women when the Asiatic Gauls were defeated by the Romans under Manlius. The centurion into whose hands she fell took advantage of his capture with a soldier's brutality and did violence to her. The man was indeed an ill-bred lout, the slave both of gain and of lust, but his love of gain prevailed; and as a considerable sum had been promised him for the woman's ransom, he brought her to a certain place to deliver her up, a river running between him and the messengers. When the Gauls crossed and after handing him the money were taking possession of Chiomara, she signed to one of them to strike the man as he was taking an affectionate leave of her. The man obeyed and cut off his head, which she took up and wrapped in the folds of her dress, and then drove off. When she came into the presence of her husband and threw the head at his feet, he was astonished and said, "Ah! my wife, it is good to keep faith." "Yes," she replied, "but it is better still that only one man who has lain with me should remain alive." Polybius tells us that he met and conversed with the lady at Sardis and admired her high spirit and intelligence.
>
> *(Plutarch, n.d., p. 465)*

Using the definitions as presented and building on the ideas of Lewin (1997) and Festinger (1957, pp. 3, 11–18, 66, 124–125, 177–181) within the context of this work, social axioms in social space, or statements assumed to be true, may now be provided:

Axiom 1. Dissonance creates psychological discomfort in a subject, leading the subject to find a means for reducing the discomfort through some combination of appropriate social action for the given situation, selective information processing, or group member support/action. Pressure to reduce dissonance in a person monotonically increases as the magnitude of his or her dissonance increases.

Axiom 2. While reducing existing dissonance, subjects will actively try to avoid social actions, information, or non-supportive group members which might cumulatively add to the existing dissonance.

Axiom 3. For a given social situation and group member position in social space, the more relevant a socially deviant event is to the subject's social position in social space, the greater the magnitude of dissonance (localized disturbance in social space) felt by the group member.

Axiom 4. For random variables R, E, and S, a given social situation, and a group member's position in social space, the greater the extent social deviation E from an accepted social norm having relevance R to the individual of interest, the greater the magnitude of the individual's social dissonance S, such that $S = R \cdot E$.

Axiom 5. For a given social situation, group member position in social space, and dissonant causing event, increasing group member social cohesion will increase the dissonance magnitude of the event for its members.

For example, in the social queue situation of a culture where one of the mores (Festinger, 1957, p. 14) or social norms is first-come, first-served, if a person external to the group intrudes ahead of a queue member already in queue, the queue member experiencing the dissonant event may try to reduce his or her dissonance by rationalizing the intrusion (i.e., maybe the queue member rationalizes that he or she is a friend of the person ahead, or that the intruder satisfactorily explained his or her need to the member who was cut in front of). If the intrusion has no reasonable explanation supporting a queue member's attempts at dissonance reduction, then one or more of the members experiencing dissonance from the intrusion may try to modify the situation by reacting against the intruder. It may also be that remaining silent is best if an action is more likely to increase dissonance above that already felt, such as when the affected queue member's personal safety is a concern if he or she were to react. In the latter case, as supported by Jarcho et al. (2011, p. 465) and Martinie et al. (2013, p. 681), making the choice of not reacting to avoid confrontation or to maintain personal safety can be a means to reduce dissonance in itself. Furthermore, as discussed generically by Festinger (1957, pp. 177–202), observation of other group member reactions influences the observing individual's probability of reaction within the group. To demonstrate this, it is shown in Chapter 3 how queue members behind and queue members ahead of a specific queue member of interest – up to the intrusion point – play a major role in determining the probability that a specific queue member reacts to an intrusion for a given social situation.

1.3 Bibb Latane's Social Impact Theory

Bibb Latane and John Darley are famous for their "Stages of Helping" analysis that assisted in understanding the then reported inaction of 38 witnesses to the murder of Katherine "Kitty" Genovese in 1964. The stages of helping Latane and Darley developed include noticing the event, interpreting the event, assuming responsibility, selecting an appropriate reaction, and then intervening. Though the number of witnesses to the Genovese murder has since been reduced through further investigation, the model stands on its own. This model, produced by Latane and Darley, has been further confirmed through experiments, most famously using subjects under various scenarios in a room filling with smoke (Latane and Darley, 1968). Results from this experiment demonstrate that subjects used other members in the room to interpret if action was necessary or not. It was found that when a subject was alone in a room filled with smoke, he or she would report the smoke 75 percent of the time within 3 minutes. If there were two passive confederates in the room with the subject, where the confederates without the subject's knowledge were instructed beforehand not to react, then the subject reported smoke only 10 percent of the time even after 6 minutes of smoke was overtly and continuously introduced into the room. We will observe later that this is similar to what happens in a queue, where queue members standing between the subject member of interest and the intruder are made use of by the subject member of interest to help interpret the situation. In particular, if queue members standing between an intruder and the subject queue member are passive (buffers as termed by Milgram), then the subject queue member is much less likely to react, even in a high-importance limited resource queue. Thirteen years later, Latane (1981) provides a second and more critical paper addressing social impact theory. In this paper, he establishes the concept which ultimately leads to the development of the social queue model introduced here. It is at this point, after introducing his results in social impact theory, that mathematical development of a social queue model will begin.

The theory of social impact addresses how other individuals for a given situation may affect the behavior of a subject member of interest. In an example, Latane (1981, p. 345, Fig. 2C) uses Milgram's data on the drawing power of crowds (Milgram et al., 1969). In this experiment, Milgram had student confederates of different size groups stand on a New York City sidewalk while looking up at a building. At the same time, passers-by were being observed and recorded by experimenters in the building to determine, post experiment, the proportion of those that looked up as a function of confederate group size. The article noted experiment anomalies, plotted results, did an analysis of variance, and then summarized the results. In 1981, Bibb Latane incorporated data from this and other experiments using his version of Stanley Stevens' power law, as proposed in 1957. His intent was to demonstrate how a power-law-like function relates to the impact others have on the individual in the same social situation. In Milgram's drawing power of crowds' experiment, Latane proposed a power function to represent the percentage of passersby looking up versus the number of confederates in a group standing and continuously looking up from a city sidewalk. Instead of deriving a density and cumulative distribution function for the data, Latane (1981) appears to have decided to be representative only in his use of the power function. This important factor aside, what he did in this article was a brilliant step in the right direction. Before continuing with this story and its analysis though, we need to go back to the nineteenth-century development of psychophysics by Gustav Fechner and then leap forward to Stanley Stevens' work of the mid-twentieth century and his power law as used by Latane.

1.4 Weber's Law and Fechner's Law

Psychophysics has been a contentious field of study since Gustav Fechner derived what is now called Fechner's Law in the mid-nineteenth century. Prior to Gustav Fechner's derivation, Ernst Weber, through experimentation during the early nineteenth century, was able to demonstrate that as the stimulus intensity I increases for any stimulus that may be sensed by one or more of the five senses, the change in stimulus intensity ΔI required for a person to just notice the difference is proportional to the reference intensity. This difference is termed the just noticeable difference (JND). With the proportionality constant (k), known as Weber's fraction, Weber's law is stated as:

$$\Delta I = k \cdot I \text{ or equivalently } \frac{\Delta I}{I} = k.$$

Consistent with Weber's law, Fechner was able to attain his goal of establishing a relationship between sensation magnitude s and stimulus intensity by assuming for any stimulus I the associated JND results in a constant value for the associated change of sensation magnitude Δs. One of only three possible functions (logarithmic, linear, or exponential) could support a constant Δs for each change in stimulus intensity resulting in a JND, and that was the logarithmic function. From this Fechner derived what is now termed Fechner's Law (Chaudhuri, 2011, pp. 9–13; Gescheider, 1997, Loc. 263),

$$s = K \cdot \log(I).$$

Note that K as used by Fechner is a constant, but not considered the same constant as the k in Weber's law. A more general form (Ekman, 1964; Dzhafarov and Colonius, 2011, p. 7; Holman and Marley, 1974, p. 197) introduces constants a and x_0. The meaning of x_0 will be developed shortly and used throughout the remainder of this chapter as a necessary means to incorporate what will be termed social noise.

Sensation magnitude $s = a + K \cdot \log(I + x_0)$.

Framing this as a sensation magnitude difference equation (Richardson, 1954), we have

$$\Delta s_n = s_{n+1} - s_n = K \cdot \log\left(\frac{I_{n+1} + x_0}{I_n + x_0}\right). \qquad\qquad \textit{Equation 1.1}$$

These are the traditional statements of Weber's Law and Fechner's Law, but there are known problems with Weber's Law at intensity values near absolute threshold (Gescheider, 1997, Loc. 93), where absolute threshold is defined as the smallest amount of stimulus intensity necessary to produce a sensation. What is observed, using Weber's Law as stated, is that $\frac{\Delta I}{I}$ is not constant when near absolute threshold as Weber's Law would predict, but instead is a monotonic decreasing function as stimulus intensity increases above absolute threshold as shown in Figure 1.1. What is observed in Figure 1.1, instead of being constant, is a function that quickly converges toward a constant value as stimulus intensity increases above absolute threshold. Likewise, Fechner's Law has theoretical implications, but until now, little apparent practical application. To really understand the development of both, and then begin to apply Fechner's Law in a form consistent with Equation 1.1 to

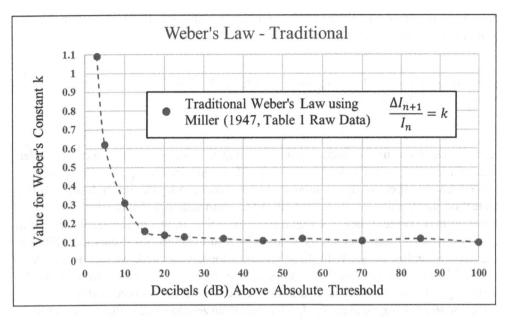

FIGURE 1.1 Traditional Form of Weber's Law Demonstrating the Nonconstant Error Near Absolute Threshold Using Miller (1947) Raw Acoustic Data.

real-world applications, we will derive Weber's Law and then Fechner's Law from scratch using the well-understood and successfully applied acoustic sonar equation.

1.5 Weber's Law Revisited

Gescheider (1997, loc. 161) indicates a modification of Weber's Law which more closely corresponds to empirical data and is given by:

$$\frac{\Delta\varphi}{\varphi + a} = k, \text{ with } a \text{ and } k \text{ constant, and } \varphi \text{ is stimulous intensity.}$$

In the same paragraph, he notes that the significance of a has not been determined but that it may represent sensory noise. In this section, using existing nomenclature within this book by replacing φ with stimulus intensity I and a with noise N, a theoretical derivation is produced showing why for an actual stimulus intensity I plus some potentially fluctuating sensory noise with intensity value N, Gescheider (1997, loc. 161) is in fact correct in his postulation. In other words, if I is replaced by $I + N$ in Weber's Law, then it will be shown by example that Weber's Law is correct for all stimulus intensity values such that

$$\frac{\Delta I}{I + N} = k, \text{ with } k \text{ constant and mean value } N.$$

To explain why Gescheider is correct, consider the passive sonar equation from underwater acoustics. This is a well-developed physical science problem and pertains directly to the problem Miller (1947, pg. 612) encountered. The passive sonar equation in decibels is given by

$$10 \cdot log\left(\frac{I}{I_{ref}}\right) - 10 \cdot log\left(\frac{N}{N_{ref}}\right) = TL + DT - DI.$$

(Urick, 1983, p.22)

Let I_{ref} represent unit reference stimulus intensity, and N_{ref} represent unit reference noise intensity such that $I_{ref} = N_{ref}$. Transmission loss (TL), in decibels, is the reduction in specific signal intensity as the signal propagates through the medium. Detection threshold (DT) in decibels is the minimum signal intensity at which an observer may correctly detect a signal from the noise. In this example, 50 percent of the time will be used. Directivity index (DI) in decibels represents the ability to reduce surrounding noise. With these variables loosely defined, we can now address the problem encountered in the traditional form of Weber's Law – that of typically assuming stimulus intensity is separate from the noise it is observed within.

As the stimulus signal intensity I_0 increases from zero to some intensity I_1 within the existing background noise, it will be detected 50 percent of the time at an intensity level above the surrounding noise floor for a given detection threshold DT. Therefore, what the operator hears when listening for the signal is not just the signal intensity I but signal intensity plus noise intensity or $I + N$. Broadband detection threshold as related to the experiment by Miller (1947) is a function of probability of detection p_d and probability of false alert p_{fa}, both of which are needed to define the detection index d, bandwidth B, and duration T of the target signal pulse. Hence, establishing broadband detection threshold, or any detection threshold for that matter, is independent of signal and noise intensity. Though adjustable for any detection probability, some example approximations for broadband detection threshold for $p_d = 0.5$ include

$$DT = 5 \cdot log\left(d\right) - 5 \cdot log\left(T \cdot B\right)$$

(Burdic, 1984, p.417),

or using a more recent modeling approach

$$DT = 10 \cdot log_{10}\left[erfc^{-1}\left(2 \cdot p_{fa}\right)\right] - 5 \cdot log_{10}\left(T \cdot B\right)$$

(Ainslie, 2010, p.597).

Given the laboratory setup of Miller's experiment, we can let $TL + DT - DI = c^*$, where c^* is a sample mean for components based on stimulus and environment (situation). The passive sonar equation using signal plus noise may now be written as

$$10 \cdot log_{10}\left(\frac{I + N}{unit\ of\ reference}\right) - 10 \cdot log_{10}\left(\frac{N}{unit\ of\ reference}\right)$$

$$= 10 \cdot log_{10}\left(\frac{I + N}{N}\right) = c^*.$$

Where *"unit of reference"* is the same for both $I + N$ and N, and recalling the relation,

$$for\ x, y \in \mathbb{R}^+,\ log(x) - log(y) = log\left(\frac{x}{y}\right).$$

Further recalling that $10^{log_{10}(x)} = x$, the passive sonar equation may be converted from decibels to stimulus intensity for comparison with Weber's Law. This conversion results in

$$\frac{I + N}{N} = 10^{\left(c^{\cdot}/10\right)} > 1\ for\ both\ I > 0\ and\ N > 0.$$

Setting $I_0 = 0$, let I_1 represent the absolute intensity threshold for base noise intensity N_0. Define $\Delta I_1 = I_1 + N_0 - (I_0 + N_0) = I_1 - 0$. Let the initial signal intensity added to the background noise intensity, given by $I_1 + N_0$, be detectable 50 percent of the time by an operator for some acceptable false alarm rate, then

$$\frac{\Delta I_1}{N_0} + 1 = \frac{\Delta I_1 + N_0}{N_0} = \frac{I_1 + N_0}{N_0} = 10^{\left(c^{\cdot}/10\right)}.$$

Let us now maintain the intensity level at I_1 and find the value I_2 such that $\Delta I_2 = I_2 - I_1$ where $\Delta I_2 + N_1$ is detectable 50 percent of the time above the newly established noise floor $N_1 = I_1 + N_0$. From this we have

$$\frac{\Delta I_2 + N_1}{N_1} = \frac{I_2 - I_1 + I_1 + N_0}{I_1 + N_0} = \frac{I_2 + N_0}{I_1 + N_0} = 10^{\left(c^{\cdot}/10\right)}.$$

The sample mean c as applied here acknowledges the new noise floor $N_1 = I_1 + N_0$ may not be distributed in exactly the same way as the base noise floor N_0 since I_1 may provide a more localized noise spectrum density. To demonstrate the progression one iteration further, let $\Delta I_3 = I_3 - I_2$ where $\Delta I_3 + N_2$ is detectable 50 percent of the time above the noise floor which is now $N_2 = I_2 + N_0$. Then

$$\frac{\Delta I_3 + N_2}{N_2} = \frac{I_3 + N_0}{I_2 + N_0} = 10^{\left(c^{\cdot}/10\right)},$$

or in general $\dfrac{\Delta I_n + N_{n-1}}{N_{n-1}} = \dfrac{I_n + N_0}{I_{n-1} + N_0} = 10^{\left(c^{\cdot}/10\right)}$ *Equation* 1.2

where $\Delta I_n = I_n - I_{n-1}$ *and* $N_{n-1} = I_{n-1} + N_0$ *for integer* $n \geq 2$.

Using $10^{\left(c/10\right)}$, a mean value, we can now solve for the more accurate version of Weber's Law:

$$\frac{\Delta I_n + N_{n-1}}{N_{n-1}} = \frac{\Delta I_n}{N_{n-1}} + 1 = 10^{\left(c/10\right)} \ or$$

$$\frac{\Delta I_n}{N_{n-1}} = 10^{\left(c/10\right)} - 1 \ for \ integer \ n \geq 2.$$

For some positive real value $k = 10^{\left(c/10\right)} - 1,$

$$10^{\left(c/10\right)} - 1 = \frac{\Delta I_n}{N_{n-1}} = \frac{\Delta I_n}{I_{n-1} + N_0} = k \ for \ integer \ n \geq 2.$$

This is the modified and empirically accurate version of Weber's Law discussed in Gescheider (1997, loc. 161), Engen (1971, pg. 18), and Stevens (1957, p. 173) which results in the ratio being constant (or nearly so since sample means are still random variables) as intended instead of monotonic decreasing when near absolute threshold.

The general relationship of the Modified Weber's Law for all values of I is then

$$k = \frac{\Delta I_n}{N_{n-1}} = \frac{\Delta I_{n-1}}{N_{n-2}} \ where \ N_{n-1} = I_{n-1} + N_0 \ for \ n \geq 2 \qquad \textit{Equation 1.3}$$

Going back to Miller (1947), Miller used a white-noise generator to produce a relatively uniform $\left(\mp 5db\right)$ broadband noise ranging from 150 to 7000 Hz. The absolute threshold for the noise intensity corresponds to 10 dB (Miller, 1947, pp. 609–612). The noise was then intermittently increased to various intensity levels for 1.5 seconds and then reduced to the original sensation intensity level $\left(I_1 + N_0\right)$ multiple times until adequate data was collected to determine the signal level $10 \cdot log_{10}\left(I_\Delta + N_0\right)$ above $\left(I_1 + N_0\right)$ at which probability of correct detection of the signal increase is 50 percent for some acceptable false alarm rate. Once the intensity level above the $I_1 + N_0$ was determined that met the detection criteria, then $I_2 + N_0$ became the next noise floor and the process was repeated. In his experiment, Miller (1947, p. 612) derived the Weber fraction value $k = 0.099$ using the traditional form of Weber's Law. Under the traditional form, this corresponds to

$$\frac{I_\Delta - I_n}{I_n} = \frac{I_\Delta}{I_n} - 1 = 0.099, \ so \ 10 \cdot log_{10}\left(\frac{I_\Delta}{I_n}\right) = 10 \cdot log_{10}\left(1.099\right) = 0.41.$$

The value 0.41 being the same value as quoted in Miller (1947, p. 612) for the highest intensities. Data from Miller (1947) as it pertains to this discussion is provided in Table 1.1.

TABLE 1.1 Results in Decibels (dB) from Miller (1947, p. 612, Table 1) – Sensation Level and Random Noise SM Columns Raw Data

$n =$	1	2	3	4	5	6	7	8	9	10	11	12
$10 \cdot log_{10}(I_n) =$	3 db	5	10	15	20	25	35	45	55	70	85	100
$10 \cdot log_{10}\left(\dfrac{I_\Delta}{I_n}\right) =$	3.2 db	2.1	1.17	.66	.55	.54	.50	.44	.50	.47	.48	.40

Using the Weber-fraction $k = 0.099$ with the Modified Weber's Law of Equation 1.3,

$$\frac{\Delta I_{n+1}}{N_n} = \frac{I_\Delta - I_n}{I_n + N_0} = k = 0.099 \ or \ \frac{I_\Delta}{I_n} = 0.099 \cdot \left(1 + \frac{N_0}{I_n}\right) + 1. \qquad \textit{Equation 1.4}$$

As sensation intensity increases above the absolute threshold noise floor, uniformity of the noise may or may not change. To avoid this question while not going too far up the sensation level curve, values for $n = 2$ are used from Table 1.1 to calculate N_0.

$$Given \ 10 \cdot log_{10}\left(\frac{I_\Delta}{I_2}\right) = 2.1, \ then \ I_\Delta = I_2 \cdot 10^{\left(2.1/10\right)} \ with \ I_2 = 10^{\left(5/10\right)}.$$

Substituting this into Equation 1.4 to solve for N_0 and canceling out I_2 results in

$$\frac{I_\Delta}{I_2} - 1 = \frac{I_2 \cdot 10^{\left(2.1/10\right)} - I_2}{I_2} = 0.6218 = 0.099 \cdot \left(1 + \frac{N_0}{10^{\left(5/10\right)}}\right) \ leading \ to$$

$$N_0 = 5.2808 \cdot I_2 = 5.2808 \cdot 10^{\left(5/10\right)} = 16.7 \ units \ of \ intensity.$$

To show Equation 1.3 represents Weber's fraction for all intensities while acknowledging sample mean variations and fluctuations due to noise, we must show

$$\frac{I_\Delta - I_n}{I_n + N_0} = \frac{\dfrac{I_\Delta}{I_n} - 1}{1 + \dfrac{16.7}{I_n}} \cong 0.099.$$

TABLE 1.2 Traditional and Modified Weber's Law Using Miller (1947, p. 612, Table 1 Raw Data) Empirical Results for Comparison with Gescheider (1997, Figure 1.4)

$n =$	1	2	3	4	5	6	7	8	9	10	11	12
$10 \cdot log_{10}(I_n) =$	3	5	10	15	20	25	35	45	55	70	85	100
$10 \cdot log_{10}\left(\dfrac{I_\Delta}{I_n}\right) =$	3.2	2.1	1.17	.66	.55	.54	.50	.44	.50	.47	.48	.40
$\dfrac{\Delta I_{n+1}}{I_n} \equiv \dfrac{I_\Delta}{I_n} - 1 =$	1.09	.62	.31	.16	.14	.13	.12	.11	.12	.11	.12	.10
$\dfrac{\Delta I_{n+1}}{N_n} \equiv \dfrac{I_\Delta - I_n}{I_n + N_0}$	.116	.099	.116	.107	.116	.126	.121	.107	.122	.114	.117	.096

$$Using\ n = 1\ as\ an\ example,\ \frac{10^{0.32} - 1}{1 + \dfrac{16.7}{10^{0.3}}} = 0.116.$$

Values, reflecting Gescheider (1997, Figure 1.4), which transform Miller's data contained in Table 1.1 into the traditional Weber's Law $\Delta I_{n+1}/I_n$, are given in Table 1.2. The last row of Table 1.2 uses Equation 1.4 with results based on the relation

$$I_\Delta = I_n \cdot 10^{log_{10}\left(\frac{I_\Delta}{I_n}\right)} and\ I_n = 10^{log_{10}(I_n)}.$$

Figure 1.2 compares the Gescheider (1997, Figure 1.4) transformation of Miller's data using the traditional form of Weber's Law $\frac{\Delta I_{n+1}}{I_n} = k$ against values derived using Equation 1.4, $\frac{\Delta I_{n+1}}{I_n + N_0} = k$.

What led to the results in Miller (1947) and independent work by Engen (1971, p. 17) was that noise was not added to the reference intensity value in the denominator as developed here. In Miller's case, the value for N was from external and internal operator intrinsic noise. In Engen's experiment, the noise was in the form of weight of the body part and associated muscles along with operator intrinsic internal noise. As will be shown, in general, noise is not limited to just physical properties – it is related to social properties as well. One example of social noise is uncertainty in how to interpret the social situation being observed due to social ambiguity resulting from possible alternate interpretations. This concept is important to psychophysics, but, more importantly, it is a critical point to understand before further developing the modeling approach within this book. We will end this chapter with a general derivation of Fechner's Law using what has been developed

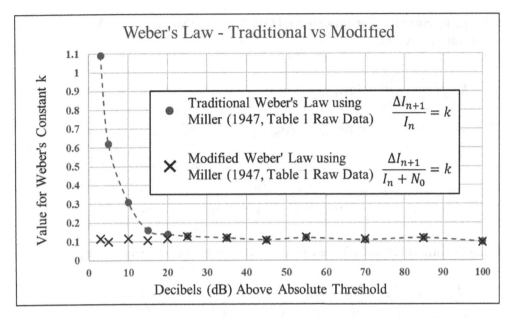

FIGURE 1.2 Modified Weber's Law Versus Results Compared to Gescheider (1997, Figure 1.4).

so far and then briefly touch on an important and more recent approach in psychophysics, the power law.

1.6 Fechner's Law Revisited

Given the derivation of the Modified Weber's Law, we now consider Fechner's Law, which, though not necessarily dependent on Weber's Law as indicated by Dzhafarov and Colonius (2011, p. 2), must be consistent with it. Using Equation 1.3 and adding 1 to both sides,

$$\frac{\Delta I_{n+1}}{N_n} + 1 = \frac{I_{n+1} + N_0}{I_n + N_0} = 10^{\left(\frac{c}{10}\right)} = k + 1 \; for \; n \geq 0 \; and \; I_0 = 0.$$

The logarithmic function has already been shown to be the necessary function, but let's go through a simple derivation and assume we are starting with an unknown function f. Label the term $\Delta s_{n,m}$ as *subjective dissimilarity*. Making use of Equations 1.1 and 1.2 with $K \in \mathbb{R}^+$, we begin with

$$subjective \; dissimilarity \; \Delta s_{n,m} = K \cdot f\left(\frac{I_n + N_0}{I_m + N_0}\right) \; for \; n \geq m, \; and \; n, m \in \mathbb{Z}^0.$$

As demonstrated in Equation 1.2, and postulated by Fechner (Dzhafarov and Colonius, 2011, p. 5; Levine and Shefner, 1991, pp. 18–19), subjective dissimilarity is constant between $I_n + N_0$ and $I_{n-1} + N_0$ when both are separated by a JND. From this, we can state in a manner similar to Shepard (1981, p. 42), that $\Delta s_{n,n-1} = \Delta s_{n-1,n-2} = C$, *a constant*. Therefore,

$$K \cdot f\left(\frac{I_n + N_0}{I_m + N_0}\right) = \Delta s_{n,m}$$

$$= \Delta s_{n,n-1} + \Delta s_{n-1,n-2} + \cdots + \Delta s_{m+1,m}$$

$$= K \cdot f\left(\frac{I_n + N_0}{I_{n-1} + N_0}\right) + \cdots + K \cdot f\left(\frac{I_{m+1} + N_0}{I_m + N_0}\right)$$

$$= (n-m) \cdot K \cdot f(k+1) \ using \ Equation \ 1.2.$$

Using Equations 1.2 and 1.3, we can also show for $n, m \in \mathbb{Z}^0$, and $n \geq m$,

$$\frac{I_n + N_0}{I_{n-1} + N_0} \cdot \frac{I_{n-1} + N_0}{I_{n-2} + N_0} \cdots \frac{I_{m+1} + N_0}{I_m + N_0} = \frac{I_n + N_0}{I_m + N_0} = (k+1)^{n-m}.$$

Combining these two results leads to

$$K \cdot f\left[(k+1)^{n-m}\right] = K \cdot f\left(\frac{I_n + N_0}{I_{n-1} + N_0} \cdot \frac{I_{n-1} + N_0}{I_{n-2} + N_0} \cdots \frac{I_{m+1} + N_0}{I_m + N_0}\right)$$

$$= K \cdot \sum_{i=m}^{n-1} f\left(\frac{I_{i+1} + N_0}{I_i + N_0}\right)$$

$$= (n-m) \cdot K \cdot f(k+1) = \Delta s_{n,m},$$

which is linearly increasing as $(n-m)$ *increases.*

As determined by Fechner and recreated by Dzhafarov and Colonius (2011, pp. 1–7), the functional relation *f* satisfies the Third Cauchy Equation, also termed the logarithmic Cauchy functional equation (Efthimiou, 2010, p. 84). Hence,

$$\Delta s_{n,m} = K \cdot log \frac{I_n + N_0}{I_m + N_0} = (n-m) \cdot K \cdot log(k+1)$$

$$m \leq n \ and \ I_m < I_n \ if \ 0 < \Delta s_{n,m}$$

where $n, m \in \mathbb{Z}^0, m \geq 1, and \ K \in \mathbb{R}^+.$ *Equation 1.5*

Equation 1.5, equivalent to the subjective magnitude discussed in Shepard (1981, p. 48) if $\alpha = 0$ and to Equation 1.1, is a generalized representation of Fechner's

Law that is used throughout the remainder of this effort. Basically, when $I_m = 0$, Equation 1.5 is indicating the subjective magnitude an observer senses when comparing expected normal background behavior N_0 with some level of socially deviant behavior $I_n + N_0$. For any fixed $m < n$, Equation 1.5 is an increasing function as n increases. Fechner's Law, as derived here and alternatively developed by Dzhafarov and Colonius (2011, pp. 3, 6), does not require reliance on JNDs. To demonstrate this, consider the following theorem and its corollary.

Theorem 1.1: Define $I_n - I_{n-1} = JND$. Given a real number p_n where $I_{n-1} < p_n \leq I_n$, $n \geq 3$

$$for\ \Delta s_{n,n-1} = K \cdot log \frac{I_n + N_0}{I_{n-1} + N_0} = K \cdot log(k+1)$$

there exists p_{n-1} where $I_{n-2} < p_{n-1} \leq I_{n-1}$ such that

$$\Delta s_{n,n-1} = K \cdot log \frac{p_n + N_0}{p_{n-1} + N_0} = K \cdot log(k+1).$$

Proof: A direct proof will be used. Based on the Modified Weber's Law, adding 1 to each side of Equation 1.3 leads to

$$\frac{\Delta I_n}{N_{n-1}} + 1 = \frac{I_n + N_0}{I_{n-1} + N_0} = k+1 \ so\ that\ I_n = (k+1) \cdot (I_{n-1} + N_0) - N_0.$$

Select any p_{n-1} where $I_{n-2} < p_{n-1} \leq I_{n-1}$, and let $p_n = (k+1) \cdot (p_{n-1} + N_0) - N_0$ noting that it is clear that $I_{n-1} < p_n \leq I_n$. Substituting these two values into Equation 1.5 leads to

$$K \cdot log \frac{p_n + N_0}{p_{n-1} + N_0} = K \cdot log \left(\frac{(k+1) \cdot (p_{n-1} + N_0)}{p_{n-1} + N_0} \right)$$

$$= K \cdot log(k+1) = \Delta s_{n,n-1}. \qquad \blacksquare$$

Corollary 1.1: Let $p \in \mathbb{R}$ such that $0 \leq p \leq I_n$, $n \in \mathbb{Z}^+$. Then,

$$K \cdot log \frac{p + N_0}{N_0} = K \cdot n \cdot log(k+1) - K \cdot log \frac{I_n + N_0}{p + N_0}$$

$$= \Delta s_{n,0} - K \cdot log \frac{I_n + N_0}{p + N_0}.$$

Proof: As a simple direct proof,

$$K \cdot log \frac{I_n + N_0}{N_0} - K \cdot log \frac{I_n + N_0}{p + N_0}$$

$$= K \cdot log \frac{p + N_0}{N_0} = K \cdot n \cdot log (k+1) - K \cdot log \frac{I_n + N_0}{p + N_0}$$

$$= \Delta s_{n,0} - K \cdot log \frac{I_n + N_0}{p + N_0}. \qquad \blacksquare$$

The implication of Theorem 1.1 and Corollary 1.1 is that the general stimulus intensity value p can exist between JNDs without loss of generality or application.

In 1957, the same year that Leon Festinger published his theory of cognitive dissonance, Stanley Stevens wrote an article (Stevens, 1957) that reinvigorated the field of psychophysics through the introduction of experimental procedures termed scaling, allowing the experimenter to directly relate stimulus intensity and subject sensation. Out of this work, the power law theory was proposed – to either build upon or replace Fechner's Law of the previous century. If s is the sensation magnitude experienced by the subject, k the scaling constant (similar in function to Weber's fraction), I the stimulus intensity, and b the stimulus specific exponent, then the power law is given by (Chaudhuri, 2011, pp. 13–14; Gescheider, 1997, Loc. 5110):

$$s = k \cdot I^b.$$

It is worth noting Ekman's (1959) observation that inserting a constant c, which must be experimentally determined along with the exponent b, improves the power law's accuracy for small values I so that

$$s = k \cdot (I + c)^b.$$

If the value for c is associated with noise N, then this begins to look familiar. In his article, Stevens' (1957) basic thesis was to demonstrate that there is a general psychophysical law relating an individual's subjective response magnitude to a stimulus magnitude and that this law indicates that equal stimulus ratios produce equal sensation ratios (Stevens, 1957, p. 162). It is not my intent, nor is it my place due to lack of expertise, to wade further into the morass of articles and arguments that have resulted within the academic community since, but one article in particular is worth noting before going to Chapter 2 where a relationship between Fechner's Law and Stevens' power law is provided.

Ekman (1964) wrote a one-page article to try and focus on the growing controversy over whether the power law is the new "true" psychophysical law or whether it is just another version of Fechner's Law. As he demonstrated mathematically, if Fechner's Law is used to describe the relation between two dissimilar stimuli, then combined they can be represented by the power law. Otherwise, Ekman (1964) argues, if used to describe the relationship between the sensation magnitude variable and a stimulus variable only, the power law is the "true" psychophysical law. Ekman's final advise was that instead of trying to solve the controversy immediately, the community needed to learn more through experimentation and model development (Fagot, 1975) before making any final determinations. At the writing of this book, and given recent literature, it would seem that the discussion has been ongoing since.

References

Allport, G. (1979). *The Nature of Prejudice* (25th Anniversary ed.). Perseus Books Publishing. https://doi.org/10.2307/3791349

Becker, H. (1934). Culture Case Study and Ideal-Typical Method: With Special Reference to Max Weber. *Social Forces*, *12*(3), 399–405. https://doi.org/10.2307/2569931

Brauer, M., & Chekroun, P. (2005). The Relationship Between Perceived Violation of Social Norms and Social Control: Situational Factors Influencing the Reaction to Deviance. *Journal of Applied Social Psychology*, *35*(7), 1519–1539. https://doi.org/10.1111/j.1559-1816.2005.tb02182.x

Burnes, B., & Cooke, B. (2013). Kurt Lewin's Field Theory: A Review and Re-Evaluation. *International Journal of Management Reviews*, *15*, 408–425. https://doi.org/10.1111/j.1468-2370.2012.00348.x

Chaudhuri, A. (2011). *Fundamentals of Sensory Perception*. Oxford University Press.

Cooper, J. (2007). *Cognitive Dissonance – Fifty Years of Classic Theory*. Sage Publications. https://dx.doi.org/10.4135/9781446214282

Dzhafarov, E.N., & Colonius, H. (2011). The Fechnerian Idea. *The American Journal of Psychology*, *124*(2), 127–140. https://doi.org/10.5406/amerjpsyc.124.2.0127

Efthimiou, C. (2010). *Introduction to Functional Equations*. University of Central Florida. Retrieved from www.msri.org/people/staff/levy/files/MCL/Efthimiou/100914book.pdf

Ekman, G. (1959). Weber's Law and Related Functions. *The Journal of Psychology*, *47*, 343–352. https://doi.org/10.1080/00223980.1959.9916336

Ekman, G. (1964). Is the Power Law a Special Case of Fechner's Law? *Perceptual and Motor Skills*, *19*(3), 730. https://doi.org/10.2466/pms.1964.19.3.730

Engen, T. (1971). Psychophysics: Discrimination and Detection. In J.W. Kling & L.A. Riggs (Eds.), *Experimental Psychology* (3rd ed., pp. 11–46). Holt, Rinehart, and Winston, Inc.

Fagot, R. (1975). A Note on the Form of the Psychophysical Function Near Threshold. *Bulletin of the Psychonomic Society*, *6*(6), 665–667. https://doi.org/10.3758/BF03337601

Festinger, L. (1957). *A Theory of Cognitive Dissonance*. Stanford University Press.

Gandhi, A. (2013). Standing Still and Cutting in Line. *South Asia Multidisciplinary Academic Journal*. https://doi.org/10.4000/samaj.3519

Gescheider, G.A. (1997). *Psychophysics: The Fundamentals*. Lawrence Erlbaum Associates, Inc., Publishers (Kindle Edition). https://doi.org/10.4324/9780203774458

Gillam, G., Simmons, K., Stevenson, D., & Weiss, E. (2014). Line, line, everywhere a line: Cultural considerations for waiting-line managers. *Business Horizons*, *57*(4), 533–539. https://doi.org/10.1016/j.bushor.2014.03.004

Herstein, I.N. (1999). *Abstract Algebra*. John Wiley & Sons.

Hogg, M.A., & Terry, D.J. (2000). Social Identity and Self-Categorization Processes in Organizational Contexts. *Academy of Management Review*, *25*(1), 121–140. https://doi.org/10.2307/259266

Holman, E.W., & Marley, A.A. (1974). Stimulus and Response Measurement. In E.C. Carterette & M.P. Friedman (Eds.), *Handbook of Perception, Vol II–Psychophysical Judgement and Measurement* (pp. 173–213). Academic Press.

Jarcho, J.M., Berkman, E.T., & Lieberman, M.D. (2011). The Neural Basis of Rationalization: Cognitive Dissonance Reduction during Decision-Making. *Social Cognitive and Affective Neuroscience*, 6, 460–467. https://doi.org/10.1093/scan/nsq054

Latane, B. (1981). The Psychology of Social Impact. *American Psychologist*, 36(4), 343–356. https://doi.org/10.1037/0003-066X.36.4.343

Latane, B., & Darley, J.M. (1968). Group Inhibition of Bystander Intervention in Emergencies. *Journal of Personality and Social Psychology*, 10(3), 215–221. https://doi.org/10.1037/h0026570

Lenski, G. (2005). *Ecological-Evolutionary Theory*. Paradigm Publishers.

Levine, M.W., & Shefner, J.M. (1991). *Fundamentals of Sensation and Perception* (2nd ed.). Brooks/Cole Publishing a Division of Wadsworth, Inc.

Lewin, K. (1935/1997). Psycho-Sociological Problems of a Minority Group. In G.W. Lewin (Ed.), *Resolving Social Conflicts and Field Theory in Social Science* (pp. 106–115). American Psychological Association (Kindle Edition). Retrieved from www.apa.org/pubs/books/4318600

Lewin, K. (1939/1997). Experiments in Social Space. In G.W. Lewin (Ed.), *Resolving Social Conflicts and Field Theory in Social Science* (pp. 58–66). American Psychological Association (Kindle Edition). Retrieved from www.apa.org/pubs/books/4318600

Lewin, K. (1940/1997). Formalization and Progress in Psychology. In G.W. Lewin (Ed.), *Resolving Social Conflicts and Field Theory in Social Science* (pp. 168–189). American Psychological Association (Kindle Edition). Retrieved from www.apa.org/pubs/books/4318600

Lewin, K. (1944/1997). The Solution of a Chronic Conflict in Industry. In G.W. Lewin (Ed.), *Resolving Social Conflicts and Field Theory in Social Science* (pp. 93–105). American Psychological Association (Kindle Edition). Retrieved from www.apa.org/pubs/books/4318600

Lewin, K. (1997). *Resolving Social Conflicts and Field Theory in Social Science*. American Psychological Association (Kindle Edition). Retrieved from https://doi.org/10.1037/10269-000

Mann, L. (1969). The Waiting Line as a Social System. *American Journal of Sociology*, 75(3), 340–354. https://doi.org/10.1086/224787

Martinie, M.A., Milland, L., & Olive, T. (2013). Some Theoretical Considerations on Attitude, Arousal and Affect during Cognitive Dissonance. *Social and Personality Psychology Compass*, 7(9), 680–688. https://doi.org/10.1111/spc3.12051

McDonald, R., & Crandall, C. (2015). Social Norms and Social Influence. *Current Opinion in Behavioral Sciences*, 3, 147–151. https://doi.org/10.1016/j.cobeha.2015.04.006

Milgram, S., Bickman, L., & Berkowitz, L. (1969). Note on the Drawing Power of Crowds of Different Size. *Journal of Personality and Social Psychology*, 13(2), 79–82. http://doi.org/10.1037/h0028070

Milgram, S., Liberty, J., Toledo, R., & Wackenhut, J. (1986). Response to Intrusion into Waiting Lines. *Journal of Personality and Social Psychology*, 51(4), 683–689. https://doi.org/10.1037/0022-3514.51.4.683

Miller, G. (1947). Sensitivity to Changes in the Intensity of White Noise and Its Relation to Masking and Loudness. *The Journal of the Acoustical Society of America*, 19(4), 609–619. https://doi.org/10.1121/1.1916528

Ostrom, E. (1990). *Governing the Commons*. Cambridge University Press. https://doi.org/10.1017/CBO9781316423936

Ostrom, E. (2012). *The Future of the Commons*. The Institute for Economic Affairs. Retrieved from http://ssrn.com/abstract=2267381

Pettigrew, T.F. (1986). The Intergroup Contact Hypothesis Reconsidered. In M. Hewstone & R. Brown (Eds.), *Contact & Conflict in Intergroup Encounters* (pp. 169–195). Basil Blackwell.

Plutarch. (n.d.). The Virtuous Deeds of Women, XXII. (W.R. Paton, Trans.). In *The Complete Histories of Polybius*. Digireads.com Publishing, Neeland Media LLC (Kindle Edition), 2014.

Richardson, C. (1954). *An Introduction to the Calculus of Finite Differences*. Van Nostrand Company, Inc.

Schmitt, B., Dube, L., & Leclerc, F. (1992). Intrusions Into Waiting Lines: Does the Queue Constitute a Social System? *Journal of Personality and Social Psychology, 63*(5), 806–815. https://doi.org/10.1037/0022-3514.63.5.806

Shepard, R. (1981). Psychological Relations and the Psychophysical Scales: On the Status of "Direct" Psychophysical Measurement. *Journal of Mathematical Psychology, 24*, 21–57. https://doi.org/10.1016/0022-2496(81)90034-1

Smith, D., Eggen, M., & St. Andre, R. (2015). *A Transition to Advanced Mathematics* (8th ed.). Cengage Learning.

Sorokin, P. (1959). *Social and Cultural Mobility*. The Free Press. Retrieved from https://ia801604.us.archive.org/8/items/in.ernet.dli.2015.275737/2015.275737.Social-And_text.pdf

Stangor, C., Jhangiani, R., & Tarry, H. (2014). *Principles of Social Psychology* (1st International ed.). BCcampus. Retrieved from https://bycheung.psych.ubc.ca/SocialPsychology Text.pdf

Stevens, S.S. (1957). On the Psychophysical Law. *The Psychophysical Review, 64*(3), 153–181. https://doi.org/10.1037/h0046162

Wasserstein, R.L., & Lazar, N.A. (2016). The ASA's Statement on P-Values: Context, Process, and Purpose. *The American Statistician, 70*(2), 129–133. https://doi.org/10.1080/0003130 5.2016.1154108

2

MILGRAM'S DRAWING POWER OF CROWDS AND SOCIAL IMPACT THEORY

> *By social impact, I mean any of the great variety of changes in physiological states and subjective feelings, motives and emotions, cognitions and beliefs, values and behavior, that occur in an individual, human or animal, as a result of the real, implied, or imagined presence or actions of other individuals.*
>
> Bibb Latane (1981, p. 343) "The Psychology of Social Impact"
> (*with permission from the American Psychological Association*)

We now have the basic tools to build on Latane's (1981) Social Impact Theory and his proposed power function $I(N) = 0.46 \cdot N^{0.24}$ used to model results from Milgram et al. (1969), where, using Latane's nomenclature, N (units of intensity I as used here) represents the number of confederates in the crowd looking up at the building. Figure 2.1 contains the raw summary data used in Milgram et al. (1969), obtained from the Stanley Milgram Papers (MS 1406), Manuscripts and Archives, Yale University Library. The power function employed by Latane (1981) is depicted in Figure 2.1 as well. To analyze and understand the underlying process leading to results, instead of using Latane's proposed power function, we need to identify an actual probability density and associated cumulative distribution function (CDF) for this data. To do this, we begin with Axioms 3 and 4 provided in Chapter 1, which state the unit stimulus sensation magnitude value s of a dissonant-causing event is a function of the event's relevance to the observer and its extent of social deviation.

2.1 Sensation Magnitude Probability Mass Function

Let the level of relevance of any event be defined as an independent continuous random variable R. Furthermore, let us define the extent social deviation of an event as the independent continuous random variable E. If the two are considered separately, then we must ask what is the relevance, and what is the extent of social deviation for an event in any given social situation? The following personal experience by this author provides an example of the concept:

DOI: 10.4324/9781003325161-3

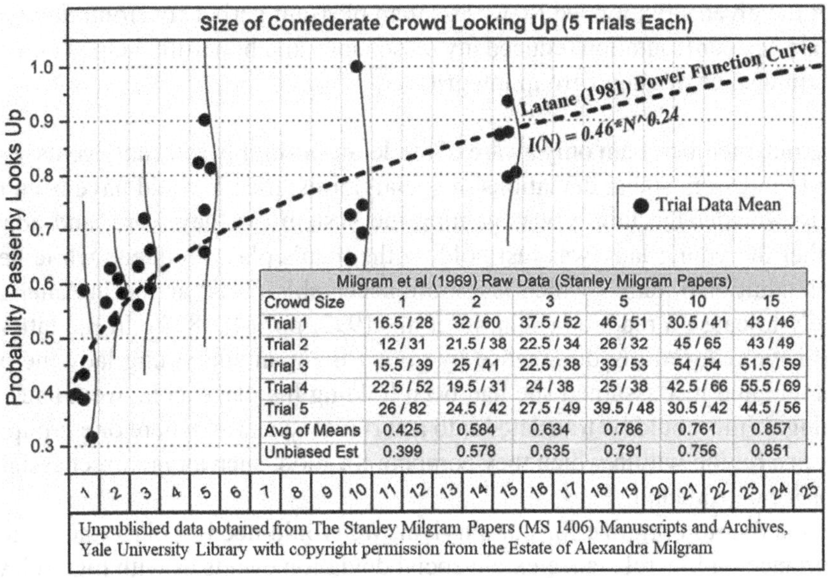
Size of Confederate Crowd Looking Up (5 Trials Each)
Probability Passerby Looks Up
1.0
0.9
0.8
0.7
0.6
0.5
0.4
0.3
Latane (1981) Power Function Curve
I(N) = 0.46*N^0.24
Trial Data Mean
Milgram et al (1969) Raw Data (Stanley Milgram Papers)
1 2 3 4 5 6 7 8 9 10 11 12 13 14 15 16 17 18 19 20 21 22 23 24 25
Unpublished data obtained from The Stanley Milgram Papers (MS 1406) Manuscripts and Archives, Yale University Library with copyright permission from the Estate of Alexandra Milgram

Crowd Size	1	2	3	5	10	15
Trial 1	16.5 / 28	32 / 60	37.5 / 52	46 / 51	30.5 / 41	43 / 46
Trial 2	12 / 31	21.5 / 38	22.5 / 34	26 / 32	45 / 65	43 / 49
Trial 3	15.5 / 39	25 / 41	22.5 / 38	39 / 53	54 / 54	51.5 / 59
Trial 4	22.5 / 52	19.5 / 31	24 / 38	25 / 38	42.5 / 66	55.5 / 69
Trial 5	26 / 82	24.5 / 42	27.5 / 49	39.5 / 48	30.5 / 42	44.5 / 56
Avg of Means	0.425	0.584	0.634	0.786	0.761	0.857
Unbiased Est	0.399	0.578	0.635	0.791	0.756	0.851

FIGURE 2.1 Latane (1981) Social Impact Power Function Compared to Milgram et al. (1969) Raw Summary Data (Stanley Milgram Papers (MS 1406), Manuscripts and Archives, Yale University Library) with Copyright Permission Obtained from the Alexandra Milgram Estate.

While in queue at a grocery store, five people were behind me and three in front. Immediately in front was a seemingly nervous teenage girl with nothing in her hands except a cell phone on which she frequently sent short texts. That she was in our group and directly in front of me made her relevant, and that she had nothing in her hands to buy made this a minor social deviation from the accepted western social norm that if waiting in queue to buy something, you should have something to buy. Both her relevance and apparent social deviation were independent of one another.

Now having my attention (Fiske and Taylor, 2017, p. 64), I noticed her furtively look two or three times in the direction of an older man a few queues over. This was again a minor deviation from accepted social norms in that typically people in one queue are not particularly concerned with observing a specific member of another queue. These two slightly dissonant events concerning the queue member were cumulative, and my interest further increased as a result.

Additional unrelated observations (i.e., background social noise) were occurring at the same time but were minor – just observations with no real significance attached. It was not until the teenage girl moved up to being next in queue to the automated interface that the full extent of the social deviation unfolding in front of me became apparent. As she moved up in position, she sent another text. The man she had looked at earlier and an older woman I had not noticed previously broke out of two nearby queues. The man, with groceries in hand, came over to join her, pushing his way past those in my queue. It was the combination of her relevance to me, and observations of minor social deviations that led to certainty of what was happening. The associated surge in extent social deviation for the given situation led to my less than pleasant verbal confrontation with the man. My confrontation reduced my dissonance and hopefully helped enforce the maintenance of queue group integrity.

If, as a group member, I am only aware of moderate-to-highly relevant events, or moderate-to-high-extent social deviations of social norms, then I would have been uncertain as to whether the family was gaming the system (as they were, and admitted) or whether the young lady was just holding the man's place in queue while he went to get an additional item – which is a more acceptable social action in America and less likely to cause a reaction (Schmitt et al., 1992, pp. 812–813). In the latter case, I would have reduced my dissonance by giving the man and young lady the benefit of the doubt and just assumed she had been holding his place as he went back to get something. If this could be translated into a survival scenario, where one group meets another in a region within which they compete for food, such an incorrect assumption could prove detrimental.

It seems that to optimize for our survival, we would need to place the same level of importance on low relevance/extent social deviation events as with high relevance/extent social deviation events. If in fact, we are somehow able to view less relevant or socially deviant observations to be as likely as highly relevant observations or socially deviant observations, then we must expect to determine and appreciate the role that information encoding (optimizing for efficiency), memory (history/experience), and attentional selection (pattern recognition) play (Fiske and Taylor, 2017, pp. 64–65) in this process. We start in this chapter with information encoding.

> **Hypothesis 2.1**: Through the evolutionary process, the human species, and likely other species, inherently apply an optimal probability mass function/distribution for both relevance and extent social norm deviation during the mental encoding process to maximize information gained from an event in social space and thus enhance chances of survival.

2.1.1 Mental Encoding and the Discrete Uniform Probability Mass Function

Let's begin by discussing how the brain distributes the probability of a two-alternative forced choice experiment, with no feedback, in the absence of information (signal). From there, we will move to the more complex example of a three-alternative forced-choice experiment without feedback to show how signal detection theory and psychophysics combine to resolve what would seem an otherwise inexplicable result in signal detection theory alone. These two examples do not offer proof that humans apply the uniform probability mass function to relevance and social deviation during normal social conditions (standard signal), but they do offer examples of sensory perception in single dimensional cases (i.e., tone, mass) resulting in nearly discrete uniform densities when only a single-type signal with no feedback is provided within an experiment requiring the subject to make one of $n \in \{2,3\}$ forced choices to identify the type of signal (i.e., loud/soft, light weight/standard weight/ heavy weight).

Prior to beginning, it is important to understand that signal detection theory addresses four conditions, correct detection of a target signal (hit), not detecting a target signal when it is present (miss), correct rejection of a signal that is not the target signal, and false classification of a non-signal as a target signal (false alarm). In the following two experiments we consider, when the standard intensity is the same as the comparative intensity, detection index equals zero, and $(n-1) \cdot p(\textit{false classification}) + p(\textit{hit}) = 1$. In essence, there is only probability of correctly identifying the target signal or incorrectly identifying the target signal (false classification – alternate choice). As a result, when standard and comparative intensities are equal (ambiguous signal), given n alternative choices, the probability of any of the n choices occurring will be shown to be uniformly distributed.

2.1.2 Tone Perception – Two-Alternative Forced Choice Experiment (No Feedback)

In an experiment which supports the prospect of a discrete uniform probability mass function under specific conditions, Tanner et al. (1967) used two amplitudes (decibel levels) of the same tone on subjects without providing feedback or information on the frequency of occurrence of the louder tone, which when correctly identified was defined as a "hit." The purpose was to compare these results, and the variation introduced with those from what was then other recent experiments conducted by independent investigators. The result of importance here was that their subjects, in a two-alternative forced choice paradigm with no feedback, perceive the probability of occurrence of the loud and soft tones to approximate a discrete uniform probability mass function when the actual probability of occurrence of the loud tone is extremely low.

Though further investigation is warranted as indicated by Hautus et al. (2022, p. 99), it seems that if there is no feedback in the two-alternative forced choice regime, subjects will act as though the schedule being used is from a [discrete] near uniform density with bias against the most frequently presented stimuli. So, when the target signal is very infrequent, the subject perceives the target signals to be nearly uniformly distributed with the nontarget signals. This is consistent with signal detection theory when the detection index is zero and with psychophysical measurement when the signals are the same (Holman and Marley, 1974, p. 197). It will be shown shortly that this can also occur in the three-alternative forced choice environment when no feedback is provided.

Viewing this in a more intuitive manner from the standpoint of social interaction, if less emphasis is given to more frequently occurring events and more emphasis to less frequently occurring events, then as a species we are geared toward more efficiently identifying behavior that is not considered normal. Otherwise, we become preoccupied or emotionally saturated with low relevance and low extent socially deviant events which, as an example, occur frequently in major cities. This may be why citizens living in dense urban environments have a greater tolerance for socially deviant behavior and focus more on their own social groups and less on others outside their social group (Milgram, 1970, pp. 1461–1462; Steblay, 1987, p. 352). From an evolutionary and possibly an information theoretic standpoint, our brains may be encoding social information into a uniform or near uniform density, which per information theory is a maximum information or maximum entropy distribution (Stone, 2015, p. 121).

On a final related note, to maximize entropy, it has been shown that optimal encoding of visual stimuli for flies and humans, having bounded neuronal output voltage (-20mV to 20mV), results in a uniform probability mass function (Stone, 2015, pp. 199–202; Laughlin, 1981). Therefore, from an evolutionary standpoint, this would seem another indication that transforming nonuniform information into uniform input may have a role in detecting and internally encoding social stimuli.

2.1.3 Mass Perception – Three-Alternative Forced Choice Experiment (No Feedback)

The design of the experiment by Tanner et al. (1967) limited a subject's choice to two options, either the tone heard was thought louder or softer. An interesting pair of related experiments were conducted by Fernberger (1913) and Arons and Irwin (1932). In both experiments, the subjects had to choose one of three alternatives. Using two subjects, Fernberger (1913) conducted an experiment using seven pair of weights of identical size placed on a rotating table. The subject was seated with his right forearm supported by a stationary table so that the hand was free to move from the wrist down. The rotating table, obscured by a screen from the subjects view, would be turned so that each weight would sequentially be placed below the hand. For each pair of weights and specified period, the subject would first lift the standard weight having a mass of 100 grams. The subject would then put down the first weight, pause a specified and consistent amount of time measured by a metronome, then lift the second weight of the pair and determine if the weight were "lighter", "equal to/doubtful", or "heavier" than the first weight. The various second weights, randomized but equally likely over the course of the experiment, had masses of 84, 88, 92, 96, 104, and 108 grams. In all, 2800 trials were performed for each of the six weight pair combinations. Table 2.1 provides the summary results.

Nineteen years after publication of Fernberger (1913), at the beginning of the Great Depression, Arons and Irwin (1932) at University of Pennsylvania made an interesting modification to Fernberger's experiment. Instead of having subjects compare various weights against a standard, some heavier and some lighter, weights were used that all fell within 20 milligrams of the standard weight of 100 grams (i.e., not a noticeable difference). As a result, in this particular experiment there were no false hits, and there

Table 2.1 Summary Data: Based on 2800 Trials of Each Weight Pair (Fernberger, 1913) with Permission from the American Psychological Association

Standard Weight = 100 grams. Second Weight =	Lighter	Equal to	Heavier
84 grams	0.9629	0.0307	0.0064
88 grams	0.8536	0.1210	0.0254
92 grams	0.6189	0.2500	0.1311
96 grams	0.3911	0.3489	0.2600
104 grams	0.0561	0.1435	0.8004
108 grams	0.0161	0.0568	0.9271

were no correct rejections. The instruction given to each subject was to judge each pair of weights and determine if the second was lighter, equal to, or heavier than the first, which in this case was the standard weight. As in Fernberger's experiment, subjects did not receive feedback on their accuracy during the trials.

Using this procedure, six subjects conducted a total of 5,500 trials, with subject results documented in sets of 50 trials. One subject performed 2,500 trials, another subject performed 1,000 trials, and the remaining four subjects performed 500 trials each. Each subject should be considered an independent random variable. Assume a Normal (Gaussian) population as done in Green and Swets (1988, p. 409), then calculate the sample standard deviation σ using all 110 sets of 50 trials from the Arons and Irwin (1932) data set. A sample mean is taken using the averaged results for each subject as provided in Table 2.2, and an approximate 95 percent confidence interval is supplied based on the combined sample standard deviation $\sigma \cong 0.095$.

Although the confidence intervals are large, Table 2.2 sample mean results indicate that when the standard mass is always introduced first, subjects tend to equalize the frequency of the three possible responses with increasing bias toward the responses "equal to" and "heavier." Figure 2.2 displays the psychometric functions based on the combined data from Fernberger (1913) and Arons and Irwin (1932), where 100 grams was used as the standard mass introduced prior to the comparative mass for each trial. The common standard deviation of $\sigma = 6.09$ used in this evaluation is the average of the "Lighter" ($\sigma_L = 6.33$) and "Heaver" ($\sigma_H = 5.85$) variances supporting a Gaussian least squares fit.

Typically, it would be expected that the "Equal" curve would cross the cumulative probability of 0.5 at 100 grams. Instead, it crosses at about 97.14 grams. This results from a known phenomenon in psychophysics called time-error, or time-order error, whose discussion of and explanation are best left to the expertise of Pratt (1933), Green and Swets (1988), Hautus et al. (2022), and Hellstrom and Rammsayer (2015). For our needs, when two stimuli of the same type and intensity are presented in sequence with some time interval between them, time-order error causes the first stimulus to seem lower in intensity than the second more recent stimulus. Figure 2.2, based on data from Fernberger (1913) and Arons and Irwin (1932) data using a least squares fit with a normal density function, demonstrates this psychophysical phenomenon with the point of subjective equality (PSE) at 97.14 grams or the weight at which the comparative stimulus appears to the subject as being equal to the standard stimulus which

TABLE 2.2 Arons and Irwin (1932, Tables III and IV) Trial Data with the Standard and Second Weight the Same. With Permission from the American Psychological Association

Both Standard and Second Weight = 100 grams	Lighter	Equal to	Heavier
Subject 1: Based on 2,500 Trials	0.263	0.316	0.422
Subject 2: Based on 1,000 Trials	0.257	0.349	0.394
Subject 3: Based on 500 Trials	0.160	0.256	0.584
Subject 4: Based on 500 Trials	0.028	0.450	0.522
Subject 5: Based on 500 Trials	0.208	0.396	0.396
Subject 6: Based on 500 Trials	0.292	0.160	0.548
Sample Mean: (Combined Sample $\sigma \cong 0.095$)	0.20	0.32]	0.48
~95% Confidence Interval (Assume Gaussian):	[0.13, 0.28]	[0.25, 0.40]	[0.40, 0.55]

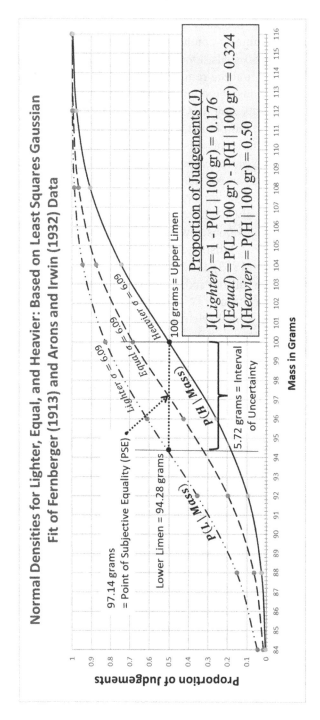

FIGURE 2.2 The Two Psychometric Functions Based on Fernberger (1913) and Arons and Irwin (1932) Data With Means at the Lower and Upper Limen Locations. The Psychometric Function for "Equal" Is Then Derived From the "Lighter" and "Heavier" Psychometric Functions While Assuming a Common Average Variance σ^2.

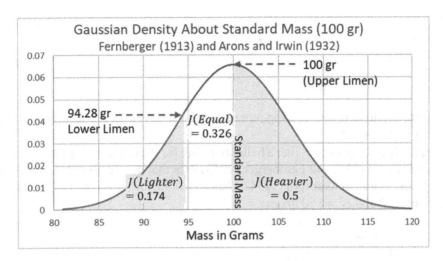

FIGURE 2.3 A More Intuitive View of Fernberger (1913) and Arons and Irwin (1932) Results Using a Gaussian Density Function with $\mu = 100$ grams (Standard Mass) and $\sigma = 6.09$.

was always presented first. The proportion of judgment results shown in Figure 2.2 are well within the 95 percent confidence intervals for each alternative as provided in Table 2.2. Using the same approach for other masses, it also shows a very close correlation with Table 2.1 sample mean results.

Based on the common standard deviation of 6.09 and relying on the Gaussian property $\Phi(-x) = 1 - \Phi(x)$ with $-\infty$ (actually 0) $< x < \infty$, Figure 2.3 provides another way in which to interpret what is occurring with the Fernberger (1913) and Arons and Irwin (1932) data. For this set of experiments, the interval of uncertainty is off center due to the standard mass of 100 grams being introduced first in every trial. It is in the interval of uncertainty that the subject perceives the standard mass and the comparative mass to be statistically equal.

A similar experiment conducted by Guilford (1954, pp. 136–137), but with only 700 trials total and an unknown number of subjects, basically removes the time-order error bias observed by Fernberger (1913) and Arons and Irwin (1932) by equally switching which mass comes first, the standard or the comparative mass. In this experiment, the standard mass is 200 grams, and the comparative masses range from 185 grams to 215 grams in steps of 5 grams. Assuming the hand is hanging down with the forearm supported as described in Fernberger (1913), the mass of my hand in such a position is approximately 150 grams – understanding the variance based on the individual and manner in which the forearm is supported exists. Let's begin with my noted hand mass as an initial noise value. To achieve psychological equivalence, with modification to account for noise, and applying the relation theory approach as discussed in Shepard (1981, p. 42), we have

$$I_n = \left(\frac{I_2 + N}{I_1 + N}\right)^{n-1} \cdot (I_1 + N) - N = (1+k)^{n-1} \cdot (I_1 + N) - N,$$

$$where\ noise\ N \sim 150\ grams, k = \frac{\left(Upper\ Limen - PSE\right)}{PSE + N},$$

$$and\ n \in \mathbb{Z}^+.$$

For a more intuitive understanding, compare this to development leading up to and including Equation 1.5.

Now, the first thing to do is find a ballpark figure for the nonnegative integer n using Arons and Irwin (1932) and Fernberger (1913) data that gets us to the standard mass of 200 grams as used in Guilford (1954). Using $I_1 = 94.28\ grams, I_2 = 97.14\ grams$ from Figure 2.2 and $N = 150\ grams$,

$$n - 1 = \frac{ln\left(\dfrac{200\ gr + 150\ gr}{94.28\ gr + 150\ gr}\right)}{ln\left(\dfrac{97.14\ gr + 150\ gr}{94.28\ gr + 150\ gr}\right)} = 30.89 \cong 31.$$

Using the Arons and Irwin (1932) and Fernberger (1913) "Lighter" data which had a Gaussian least squares fit of 0.0012, and given the "Heavier" mean mass was 100 grams, we will find values for I_2 and N such that

$$I_3 = 100\ gr = \left(\frac{I_2 + N}{94.28\ gr + N}\right)^{3-1} \cdot \left(94.28\ gr + N\right) - N,\ and$$

$$I_{32} = 200\ gr = \left(\frac{I_2 + N}{94.28\ gr + N}\right)^{32-1} \cdot \left(94.28\ gr + N\right) - N.$$

Having two equations and two unknowns, the result is,

$$I_2 = 97.12319\ grams, and\ N = 146.127\ grams.$$

The point of subjective equality (PSE) value, the original I_2 value shown in Figure 2.2, is then adjusted from 97.14 grams to 97.1232 grams. We may now find the lower and upper limen values I_{30} and I_{32} for the Guilford (1954) experiment based again on relation theory while accounting for noise such that

$$I_{31} = \left(\frac{97.12319\ gr + 146.127}{94.28\ gr + 146.127}\right)^{31-1} \cdot \left(94.28\ gr + 146.127\right)$$

$$- 146.127 = 195.96\ grams.$$

$I_{32} = 200.00 \, grams.$

$$I_{33} = \left(\frac{97.12319 \, gr + 146.127}{94.28 \, gr + 146.127} \right)^{33-1} \cdot \left(94.28 \, gr + 146.127 \right)$$
$$-146.127 = 204.09 \, grams.$$

Although the Guilford (1954) data only had 100 trials per comparative mass and an unknown number of subjects, this derivation shown in Figure 2.4 does suggest an interesting result based on the revised data resulting from relation theory. Using the values I_{31}, I_{32}, and I_{33} as the psychometric function Gaussian means, and assuming a common standard deviation $\sigma = 9.40$ based on a least squares fit of the empirical data, when the order of the standard mass is switched equally with that of the comparative mass as Guilford did in his experiment, the proportion of judgments are nearly uniform across the three alternative choices. As an aside, Weber's fraction for this set of three independent experiments is found to be 0.0238 for the transition from "lighter" to "heavier."

Guilford (1954, p. 137) evaluated the data four different ways. In one of the four methods, he made use of normal graphic probit paper. Analyzing the data in that manner, Guilford derived a Lower Limen Value of 196.2 grams with $\sigma = 9.6$, a PSE of 200.2 grams, and an Upper Limen Value of 204.3 grams with $\sigma = 10$. With the minor difference in standard deviation values, Guildford's psychometric function mean values are exactly the same as derived here if shifted 0.2 grams to the left. As with the Arons and Irwin (1932) data, Figure 2.5 provides a more intuitive way to consider the results using the Guilford (1954) data.

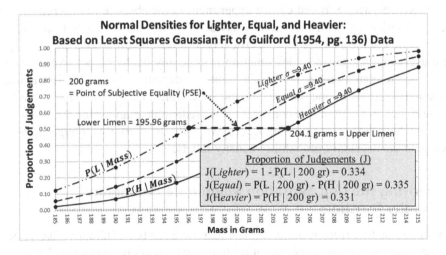

FIGURE 2.4   The Two Psychometric Functions Based on Fernberger (1913) and Arons and Irwin (1932) Upper and Lower Limen Values Using Guilford (1954, p. 136) Results to Calculate the Standard Deviation Based on a Least Squares Fit. The Psychometric Function for "Equal" is Derived from the "Lighter" and "Heavier" Psychometric Functions and Assumes the Same Variance σ^2.

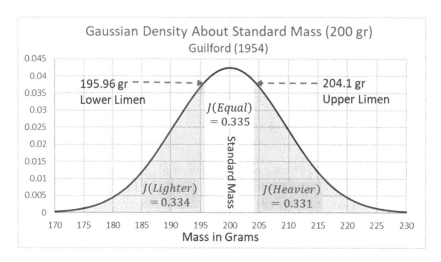

FIGURE 2.5   More Intuitive View of Guilford (1954) Results Using a Gaussian Density Function with $\mu = 200$ grams (Standard Mass) and $\sigma = 9.40$.

Consider now two or more socially deviant events of the same type and intensity occurring within a short but observable interval of time. It may be that time-order error as experienced by Fernberger (1913) and Arons and Irwin (1932) is also present when comparing the intensity of such social events. It will be seen later that this concept is subjectively addressed in cognitive dissonance theory and quantitatively modeled in Chapter 6. The preceding experiments also support the hypothesis of a uniform probability mass function in one dimension for the three-alternative forced choice paradigm when the standard and comparative signals are the same but equally switched in their presentation sequence. Tanner et al. (1967) demonstrated the same in their two-alternative forced choice experiment in the absence of ordered pairs and hence time-order error considerations. It is not out of the question then to consider the possibility that in the absence of a specific target signal and time-order in social space, where everyone in a given social situation is generally behaving per expected social norms, each human member in that situation is independently and uniformly encoding social deviations in one dimension, and social relevance in another – some more, some less, but on average in a consistent and predictable manner for the given social norms in which the social situation occurs. We will now carry this thought forward in developing the necessary probability density function.

2.2 Deriving the Unit Sensation Magnitude Probability Density Function

We assume it is true from Axioms 3 and 4 that there are two random variables generating an observer's dissonance sensation magnitude, namely relevance R and extent social deviation E of observed event(s) based on the observer's social norms and social situation within which the event(s) occur.

Proposition 2.1: The random variable representing relevance R of a social event for a given social situation is bounded below by zero, and above by one.

Proof: Relevance must be bounded below by zero since a social event is either relevant to some degree or it is not (i.e., zero relevance). Now, using proof by contradiction, assume relevance of a social event is not bounded above. Given that relevance is not bounded above, then for any event A having immediate impact on an individual or group, there is some event B that is more relevant based on impact to an individual group member for a given social situation. Let A be the termination of the individual or group through some immediate means. Since the individual or group now no longer exists, there is nothing that can have a greater impact on the individual or group, hence B cannot exist. By contradiction, relevance of a social event is bounded below by zero, and above by some finite value. Since A is an arbitrary constant, set $A = 1$.

Before moving on to Proposition 2.2, a historical account by Cassius Dio, the son of a Roman senator who lived about 1,800 years ago, should be told to make the point that the most egregious forms of social deviation all seem to fall into the dark bin of arbitrary gruesome death. It also indicates that Augustus Caesar might have been aware of some of the concepts as now contained in cognitive dissonance theory (Dio, 1917).

This same year Vedius Pollio died, a man who in general had done nothing deserving of remembrance, as he was sprung from freedmen, belonged to the knights, and had performed no brilliant deeds; but he had become very famous for his wealth and for his cruelty, so that he has even gained a place in history. Most of the things he did would be wearisome to relate, but I may mention that he kept in reservoirs huge lampreys that had been trained to eat men, and was accustomed to throw to them such of his slaves as he desired to put to death. Once, when he was entertaining Augustus, his cupbearer broke a crystal goblet, and without regard for his guest, Pollio ordered the fellow to be thrown to the lampreys. Hereupon the slave fell on his knees before Augustus and supplicated him, and Augustus at first tried to persuade Pollio not to commit such a monstrous deed. Then, when Pollio paid no heed to him, the emperor said, "Bring all the rest of the drinking vessels which are of like sort or any others of value that you possess, in order that I may use them," and when they were brought, he ordered them to be broken. When Pollio saw this, he was vexed of course; but since he was no longer angry over the one goblet, considering the great number of others that were ruined, and, on the other hand could not punish his servant for what Augustus also had done, he held his peace. This is the sort of person Pollio was, who died at this time. Among his many bequests to many persons he left to Augustus a good share of his estate together with Pausilypon, the place between Neapolis and Puteoli, with instructions that some public work of great beauty should be erected there. Augustus razed Pollio's house to the ground, on the pretext of preparing for the erection of the other structure, but really with the purpose that Pollio should have no monument in the city; and built a colonnade, inscribing on it the name, not of Pollio, but of Livia [wife of Augustus].

(pp. 339–343)

The important concept to focus on in this story, which leads us into the next proposition, is that of breaking the crystal goblets and the anxiety this caused Vedius Pollio.

Even though it is certain this caused the slave a certain amount of anxiety as well, it is Vedius Pollio's reaction that is of interest. In this situation, the first cup that is broken angers Vedius Pollio. After Augustus has the remaining cups broken, the breaking of one cup becomes minor in comparison to the cumulative effect, and Vedius Pollio can do nothing against Augustus Caesar, nor could he stop him from having the glasses broken. Additionally, if Vedius Pollio were to still desire throwing the slave to the lampreys then he might further consider what else Augustus would do in response that could be even worse (see Axiom 2, Chapter 1). As a bystander, years safely distant from the event and from Augustus, what Augustus did was clever and would indicate a keen understanding of human psychology if the story is accurate.

> **Proposition 2.2**: The extent social deviation E for a single unit social stimulus and situation is bounded. If more than one-unit social stimulus is applied simultaneously or near simultaneously, then the extent social deviation may eventually exceed existing bounds and be considered unbounded.
>
> **Proof**: (Unit Stimulus) Extent social deviation must be bounded below by zero since something is either a social deviation for a given situation, or it is not (i.e., zero extent social deviation). As a proof by contradiction, assume extent of social deviation for a unit stimulus is not bounded above. Assuming no inhibitions or uncertainty of interpretation by the group member observing the deviation, let some event A represent the greatest deviation from social norms for the given situation that a given group member can comprehend, and in which the member or group would not tolerate. Since extent social deviation is not bounded above, there must exist some event B having a greater extent social deviation than A. But since A represents the worst deviation from social norms for the given situation that a given group member can comprehend at the time, then B cannot exist since it has not manifested itself historically or through experience by any member. In other words, until B manifests itself, event A is the perceived upper bound of social deviation for a given social situation. Hence, by contradiction, social deviation for a given social situation is bounded below by zero, and above by some finite value.
>
> (Multiple Stimuli) Assume an extent deviation A occurs multiple times simultaneously. An example being a soldier seeing multiple friends being killed in war within just a few seconds as they try to take a new tactical position as ordered. Then if A represents the extent deviation of one friend being killed in action, two friends being killed in action at the same or nearly same time would exceed A. Therefore, in this case, A is not bounded.

The first situation, that of a unit stimulus, will be addressed first since it is the most intuitive. With that done, and the theory developed, the situation with unbounded stimuli may then be addressed.

> **Proposition 2.3**: Per Axiom 4 and Propositions 2.1 and 2.2, a unit (i.e., single) stimulus sensation magnitude resulting in social dissonance experienced by an observer is bounded.

Proof: Using Propositions 2.1 and 2.2, define the bounded range of the random variables R *and* E by $0 \leq R \leq 1$ and $0 \leq E \leq d$ *with* $d \in \mathbb{R}^+$. Consider events that are socially dissonant with some density based on the random variable for unit stimulus sensation magnitude S. Since R and E are independent, per Rohatgi (1976, p. 141) and Glen et al. (2004, pp. 452–453), S is bounded below by 0 and above by d such that

$$S = R \cdot E \rightarrow (0, d].$$

The random variable S is undefined at zero since that would include all observable or known events not socially dissonant (i.e., everything else). Hence, unit stimulus sensation magnitude is bounded below and above.

With the necessary propositions now in place, but prior to developing and introducing the density function for social dissonance, it is important that we visit the highly relevant work of Sherif and Sherif (1956) and of Zipf (1948).

In their discussion on the formation of group norms, Sherif and Sherif (1956, pp. 170–171) state Axioms 4 and 5 in alternative form. While assuming relevance to the group, they sum up their view by observing that toleration of a social norm deviation decreases with the norm's increasing importance. In effect, this presents an inverse relationship between tolerated behavior and dissonance sensation magnitude. Zipf (1948, loc. 10890), when providing samples of Congressional action, also determined that legal penalties are inversely related to the frequency of the crime. He more importantly ends with the observation that if high-magnitude social disturbances were frequent, they would eventually destroy the social system experiencing them. If magnitude of disturbance as used by Zipf is replaced with sensation magnitude, then combining the work of Festinger, the Sherifs', and now Zipf, we may conclude that as sensation magnitude of any relevant socially deviant event increases, so should the penalty (pressure imposed) by the group against the instigator. As a result, the unit stimulus sensation magnitude value s should be represented by a density function such that s has an inverse relationship to frequency of occurrence, given effective penalties (pressure) by the group are commensurate with the extent social deviation and relevance as perceived by group members.

With this we will prove our next theorem.

Theorem 2.1: Given Proposition 2.3, unit stimulus sensation magnitude random variable S with values $s_0, s_1 \in (0, d = 1] = R_s = \{s \mid f(s) > 0\}$ where R_s is the range of S, and the associated density function f_s, constrained by

1. $f_S(s_0) \geq f_S(s_1) \geq 0$ for $0 < s_0 < s_1 \leq d$;
2. $f_S(d) = 0$; and
3. $\int_{0^+}^{d} f_S(s) \, ds = 1$

is $f_S(s) = \dfrac{1}{d} \cdot ln\left(\dfrac{d}{s}\right)$ based on the Cauchy equation meeting these criteria.

Proof: Given the half open unit stimulus sensation magnitude value $s \in (0, d = 1]$, it must be shown that only one of the four Cauchy family of equations satisfy the three stated criteria in the Theorem. The four Cauchy equations are:

1. The Linear Cauchy functional equation: $f_S(s_0 + s_1) = f_S(s_0) + f_S(s_1)$
2. The Exponential Cauchy functional equation: $f_S(s_0 + s_1) = f_S(s_0) \cdot f_S(s_1)$
3. The Logarithmic Cauchy functional equation: $f_S(s_0 \cdot s_1) = f_S(s_0) + f_S(s_1)$
4. The Power Cauchy functional equation: $f_S(s_0 \cdot s_1) = f_S(s_0) \cdot f_S(s_1)$

Given the constraints, it may be concluded that $f_S(s_0 + d)$ is undefined under

1. *The linear Cauchy equation since $s_0 + d > d$; and*
2. *The exponetial Cauchy equation since $s_0 + d > d$.*

The remaining two possible Cauchy equations to be analyzed are:

3. *The logarithmic Cauchy equation:*
   a. It is certainly true that $f_S(s_0 \cdot s_1) \geq f_S(s_0) \geq f_S(s_1)$ since $s_0 \cdot s_1 \leq s_0 < s_1 \leq d$, making it possible for $f_S(s_0 \cdot s_1) = f_S(s_0) + f_S(s_1)$. It is also true that $f_S(s_0 \cdot d) = f_S(s_0) + f_S(d) = f_S(s_0) + 0 = f_S(s_0)$.

4. *The power Cauchy equation:*
   a. $f_S(s_0 \cdot s_1) = f_S(s_0) \cdot f_S(s_1)$ is false for $s_1 = d$ since $f_S(s_0 \cdot d) = 0$ *for all s_0*

   $$implying \int_{0^+}^{d} f_S\left(\frac{s}{d} \cdot d\right) ds = 0$$
   .

The only Cauchy equation that possibly meets the three constraints is the logarithmic Cauchy equation. It remains to find the function $f_S(s)$ such that $\int_{0^+}^{d} f_S(s) ds = 1$. This may be determined through simple integration. Since

$$a \int_{0^+}^{1} ln(s) ds = a \cdot \left[s \cdot \ln(s) - s \right]_{0^+}^{1}$$
$$= \lim_{s \to 0} \left(a \cdot \left[1 \cdot ln(1) - 1 \right] - a \cdot \left[s \cdot \ln(s) - s \right] \right) = 1.$$

Where $\lim\limits_{s \to 0} \left[-s \cdot \ln(s) \right] = 0$ is found by noting $-s \cdot \ln(s) = \dfrac{\ln\left(\frac{1}{s}\right)}{\frac{1}{s}}$. Using L'Hopital's Rule,

$$\lim_{s \to 0}\left[-s \cdot \ln(s)\right] = \lim_{s \to 0} \frac{-\dfrac{d}{ds}\ln(s)}{\dfrac{d}{ds}\dfrac{1}{s}} = \lim_{s \to 0} \frac{-\dfrac{1}{s}}{\dfrac{-1}{s^2}}$$

$$= \lim_{s \to 0} s = 0$$

Therefore, $-a = 1$, *or equivalently* $a = -1$.

Substituting -1 for the variable a, $f_S(s) = -1 \cdot \ln(s) = \ln\left(\dfrac{1}{s}\right)$ and $\int_{0^+}^{1} \ln\left(\dfrac{1}{s}\right) ds = 1$. Hence,

$$\frac{1}{d} \cdot \int_{0^+}^{d} \ln\left(\frac{d}{s}\right) ds = 1, \text{ therefore } f_S(s) = \frac{1}{d} \cdot \ln\left(\frac{d}{s}\right), 0 < s \le d. \qquad \blacksquare$$

Let us now consider the development of the density function in Theorem 2.1 from another direction by assuming from previous arguments that R (relevance) is an independent standard uniform random variable, and E (extant social deviation) is a uniform random variable such that $0 \le E \le d$. Next, assume that if a socially deviant event is not relevant to the group or groups an individual identifies with at the time, then there is no increase in dissonance to the individual. Likewise, if a relevant event occurs that does not deviate from social norms of the group or groups the individual identifies with at the time, then no increase in dissonance will occur for the individual. Making use of Rohatgi (1976, p. 141), we can now prove our second theorem.

> **Theorem 2.2**: Given independent uniformly distributed random variables R *and* E with $0 \le R \le 1$ and $0 \le E \le d$, and domain element $s \in R_S = \{s \mid f(s) > 0\}$, the probability density function associated with the random variable S, repre-
>
> senting frequency of occurrence is $f_S(s) = \dfrac{1}{d} \cdot \ln\left(\dfrac{d}{s}\right)$.

Proof: Direct proof may be performed using Pham-Gia and Turkkan (2002) with $S = \beta(1,1) \cdot \beta(1,1)$ and then converting for a nonstandard uniform density as was done in Theorem 2.1, or the more straightforward approach by Rohatgi (1976, p. 141) may be used,

$$f_S(s) = \int_{s}^{d} \frac{1}{d} \cdot \frac{1}{x} dx = \frac{1}{d} \cdot \ln\left(\frac{d}{s}\right), 0 < s \le d. \qquad \blacksquare$$

The density function $f_S(s) = \dfrac{1}{d} \cdot \ln\left(\dfrac{d}{s}\right)$ derived in Theorem 2.1 and in Theorem 2.2 supports the arguments presented by Sherif, Zipf, and Festinger.

The probability density function $f_S(s) = \dfrac{1}{d} \cdot \ln\left(\dfrac{d}{s}\right)$ and its associated CDF,

$$F_S(s) = \frac{1}{d} \cdot \int_{0^+}^{s} ln\left(\frac{d}{x}\right) dx = \frac{s}{d} \cdot \left[1 - ln\left(\frac{s}{d}\right)\right], 0 < s \leq d,$$

will be used to begin model development by converting data from Milgram et al. (1969) so as to determine $\Delta s_{n,0}$ as defined in Equation 1.5 – the change in sensation magnitude s of a passerby as a function of number of people in the crowd of confederates looking up versus no crowd. We will then consider the results and their implications with regard to information theory and the necessary shift from uniform to exponential encoding in the presence of multiple simultaneous stimuli of the same type. The chapter ends by proposing a simple relationship between Fechner's Law and the power law – demonstrating, in this application at least, that equivalence exists under exponential encoding.

2.3 Psychophysical Solution to Milgram's 1969 Drawing Power of Crowds Experiment

Information theory seems to play a key role in how external stimuli are optimally encoded for transmission and processing to the brain when observed. The derived sensation magnitude probability density function pertains to the bounded variable having bounded unit stimulus sensation magnitude $s \in (0, d]$. Given experimental data shows probability of reaction increases with crowd size, we might infer that sensation magnitude increases as well. We must therefore consider what happens from an encoding standpoint not only for a single stimulus, but when we are simultaneously exposed to a potentially unbounded number $p \in \mathbb{Z}^0$ of unit stimuli of the same type as well. Letting $s_P = \Delta s_{p,0}$ (Equation 1.5), when p units of stimuli are introduced to the observer at the same time, then $\lim_{p \to \infty} f(s_P) \to \infty$ for some function f such that $s_P \in [0, \infty)$. For the crowd-gathering experiment, this would equate to an infinite size crowd of confederates looking up at the building. Obviously, there are only a finite number of confederates that will fit on a city block while still allowing passersby, but the concept and therefore the question remain. According to information theory, such an unbounded situation results in encoding the information using an exponential distribution (Ben-Naim, 2017, pp. 105–106) to maximize Shannon's Measure of Information (of which entropy is a subset), not the two uniform distributions which led to the logarithmic sensation magnitude probability density function derived in Section 2.2 for the bounded case. In this section, we will derive probability of looking up as a function of crowd size for both cases, discuss the meaning, select one of the two cases presented along with the rationale for doing so, and move forward into modeling the queue which uses what is derived here for the foundation of a more comprehensive and interesting quantitative social model.

2.3.1 Encoding: Uniform Distribution as the Basis

Returning to Figure 2.1, after five trials and 232 observations, Milgram et al. (1969) found the probability of a passerby looking up to be 0.425 if only one confederate on the sidewalk is looking up at the building. Why the average of the five trial-means was used for each crowd configuration instead of an unbiased estimator or stratified sample mean is not critical to this discussion or the original paper, but to maintain continuity with the original paper and Latane (1981), the average of the five-trial means for each of the six crowd-gathering conditions as originally evaluated will be used. To do otherwise would imply a stratified random sampling approach which without understanding the original trial environment and experimental design would be hard to justify.

Definition 2.3: Let the random variable S apply to a single unit observation of a given socially deviant event and its relevance to an individual for the social situation in a noiseless environment. Then s_p is the sensation magnitude value when p units of the same stimulus intensity are present under uniform density encoding and an independent random variable for social noise N_0 such that

$$0 \leq s_P = S \cdot ln\left(\frac{p+N_0}{N_0}\right) \leq d, p \in \mathbb{Z}^0, and \ N_0 \in \mathbb{R}$$

$$where \ \frac{1}{e-1} \leq N_0 < \infty.$$

Equation 2.1

When n observations with the same number of units p are used each time under the same conditions (i.e., situation and stimuli), but with different individuals so as to allow a sample mean, then the term $\bar{s}_P$ is used and is defined as,

$$\bar{s}_P = \frac{1}{n} \cdot \sum_{i=1}^{n} S_i \cdot ln\left(\frac{p+N_{0,i}}{N_{0,i}}\right) constrained \ by \ \bar{s}_P \in R_S \ under \ uniform$$

encoding.

For clarification, we assume the same stimulus intensity p is observed each time by each individual, but that each individual has varying values of N_0 at the time of observation. This results in $ln\left(\frac{p+N_0}{N_0}\right)$ being a random variable since it is a function of the random variable N_0.

Definition 2.4: Define the sample mean $\bar{s} = \bar{s}_1$ for $N_0 = \frac{1}{e-1}$.

Using Definition 2.3, to map the probability of a passerby looking up (given one confederate standing on the sidewalk looking up) into density space based on the probability density function proven in Theorems 2.1 and 2.2, let's begin by setting $d = 1$ from Theorem 2.2 and take the inverse of the cumulative density function $F_S(\bar{s}_1) = 0.425$, where probability of reaction value 0.425 is the mean probability for a passerby to look up (i.e., react) if one confederate is looking up. Then

TABLE 2.3 Empirical $\bar{s}_P$ Values Using Theorem 2.1 and its Inverse CDF with d = 1. Empirical Data Derived from the Stanley Milgram Papers (MS 1406)

Crowd Condition	$p = 1$	$p = 2$	$p = 3$	$p = 5$	$p = 10$	$p = 15$
Empirical Probability $F_S\left(\bar{s}_P\right)$	0.425	0.584	0.634	0.786	0.761	0.857
Empirical $\bar{s}_P = F_S^{-1}\left[F_S\left(\bar{s}_P\right)\right]$	0.145	0.241	0.278	0.422	0.394	0.515

$$knowing\ F_S^{-1}\left[F_S\left(\bar{s}_1\right)\right] = \left[F_S^{-1}F_S\right]\left(\bar{s}_1\right)$$
$$= \bar{s}_1, we\ have, using\ d = 1,$$

$$F_S^{-1}\left(0.425\right) = \bar{s}_1 = 0.145\ so\ that\ F_S\left(\bar{s}_1\right)$$
$$= 0.145 \cdot \left[1 - ln\left(0.145\right)\right] = 0.425.$$

The remaining values for $\bar{s}_P, p > 1$, are calculated for crowd size $p \in \{2,3,5,10,15\}$ in a similar manner and provided in Table 2.3.

Equation 2.1 is in the form of Equation 1.5, supported by Corollary 1.1. For an unknown reason, the calculated values for conditions $p \in \{5,10\}$ are just at the edge of their respective 95 percent confidence intervals (Devore, 1987, p. 253) from the sample mean results (Figure 2.1). It is noted in the original typed results from Stanley Milgram Papers (MS 1406, Box 24, Folder 3) that some passersby were looking up both before and after they entered the defined observation area where data was collected. Additionally, nowhere is it stated in the remaining papers what the environmental conditions were during each observation or what the time of day was (i.e., morning or evening rush hour, lunch, etc.). Whether any of these caused the wide variations observed in conditions $p \in \{5,10\}$ is unknown, and after 50 years since the experiment was conducted, the one remaining author does not recall. As for those passersby who looked up, or stood and looked up, it is assumed for now that their impact was minimal in relation to the confederate crowd size. The data as available just does not support a more analytic argument.

To show $s_p = S \cdot ln\left(\dfrac{p + N_0}{N_0}\right)$ is an adequate model for this social situation, we must determine both the value for social noise N_0 and the sample mean unit stimulus sensation magnitude value $\bar{s}$ of Definition 2.4 based on the six trial conditions.

Since S and N_o are independent of one another, so are S and $ln\left(\dfrac{p + N_0}{N_0}\right)$. Hence, $E\left[S \cdot ln\left(\dfrac{p + N_0}{N_0}\right)\right] = E[S] \cdot E\left[ln\left(\dfrac{p + N_0}{N_0}\right)\right]$. Using the Strong Law of Large Numbers, with $\dfrac{1}{e-1} < N_0 < \infty$, there exists a geometric mean $\widehat{N}_0$ such that for N_i

independent identically distributed random variables drawn from each instance of N_0 associated with each reacting and nonreacting individual member,

$$E\left[ln\left(\frac{p+N_0}{N_0}\right)\right] = \lim_{n\to\infty}\left[\frac{1}{n}\cdot\sum_{i=1}^{n}ln\left(p+N_i\right) - \frac{1}{n}\cdot\sum_{i=1}^{n}ln\left(N_i\right)\right]$$

$$= \lim_{n\to\infty}\left[ln\left[\prod_{i=1}^{n}\left(p+N_i\right)\right]^{\frac{1}{n}} - ln\left(\widehat{N}_0\right)\right].$$

Since $E\left(N_0\right) = \frac{1}{n}\cdot\sum_{i=1}^{n}N_i \geq \left(\prod_{i=1}^{n}N_i\right)^{\frac{1}{n}}$

$= \widehat{N}_0$, *noting* $ln\left(\dfrac{p+N_0}{N_0}\right)$ *is a convex function of*

N_0, *and that by induction* $E\left[ln\left(p+N_0\right)\right] \geq ln\left(p+\widehat{N}_0\right)$, *then by applying Jensons' Inequality,*

$$E\left[ln\left(\frac{p+N_0}{N_0}\right)\right] \geq ln\left(\frac{p+\widehat{N}_0}{\widehat{N}_0}\right)$$

$$= ln\left(\frac{p}{\widehat{N}_0}+1\right)$$

$$\geq ln\left[\frac{p}{E\left(N_0\right)}+1\right] = ln\left[\frac{p+E\left(N_0\right)}{E\left(N_0\right)}\right].$$

For p, a constant in the given social situation, assume for simplification $E\left[ln\left(\dfrac{p+N_0}{N_0}\right)\right] \cong ln\left(\dfrac{p+\widehat{N}_0}{\widehat{N}_0}\right)$ since $\widehat{N}_0$ exists as the denominator of both. We can now move forward to approximate the sample mean of s, designated $\overline{s}$ per Definition 2.4, where as per Equation 2.1,

$$\overline{s}_p \cong \overline{s}\cdot ln\left(\frac{p+\widehat{N}_0}{\widehat{N}_0}\right), \text{ integer } p \geq 0 \text{ and } \frac{1}{e-1} < \widehat{N}_0 < \infty. \qquad \textit{Equation 2.2}$$

Finding the sample mean unit sensation magnitude $\bar{s}$ requires the development of three theorems before an analysis of Milgram's crowd data is possible.

Theorem 2.3: Given $\widehat{N}_0, a \in (0, \infty)$, $i, j \in \mathbb{Z}^+$, and $\bar{s} \leq d$,

$$\bar{s} \cdot ln\left(\frac{j+a}{a}\right) \cdot \frac{ln\left(\frac{i+\widehat{N}_0}{\widehat{N}_0}\right)}{ln\left(\frac{j+\widehat{N}_0}{\widehat{N}_0}\right)} = \bar{s} \cdot ln\left(\frac{i+a}{a}\right) = \bar{s}_i \text{ if and only}$$

if $\widehat{N}_0 = a$.

Proof: The proof in both directions will be a direct proof. Addressing the backward direction first, if $\widehat{N}_0 = a$ then clearly

$$\bar{s} \cdot ln\left(\frac{j+a}{a}\right) \cdot \frac{ln\left(\frac{i+a}{a}\right)}{ln\left(\frac{j+a}{a}\right)} = \bar{s} \cdot ln\left(\frac{i+a}{a}\right) = \bar{s}_i.$$

The forward direction is a little more complicated.

$$\text{If } \bar{s} \cdot ln\left(\frac{j+a}{a}\right) \cdot \frac{ln\left(\frac{i+\widehat{N}_0}{\widehat{N}_0}\right)}{ln\left(\frac{j+\widehat{N}_0}{\widehat{N}_0}\right)} = \bar{s} \cdot ln\left(\frac{i+a}{a}\right)$$

$$\text{then } \frac{ln\left(\frac{i+\widehat{N}_0}{\widehat{N}_0}\right)}{ln\left(\frac{j+\widehat{N}_0}{\widehat{N}_0}\right)} = \frac{ln\left(\frac{i+a}{a}\right)}{ln\left(\frac{j+a}{a}\right)}.$$

This is clearly true for $\widehat{N}_0 = a$. To show this is the only case in which the relation is true, consider the derivatives of the left side as a function of $\widehat{N}_0$ such that

$$\frac{d}{d\widehat{N}_0} ln\left(\frac{i+\widehat{N}_0}{\widehat{N}_0}\right) = \frac{-i}{\widehat{N}_0 \cdot (i+\widehat{N}_0)} \text{ and } \frac{d}{d\widehat{N}_0} ln\left(\frac{j+\widehat{N}_0}{\widehat{N}_0}\right)$$

$$= \frac{-j}{\widehat{N}_0 \cdot (j+\widehat{N}_0)}$$

Since

$$\lim_{\widehat{N}_0 \to \infty} ln\left(\frac{i+\widehat{N}_0}{\widehat{N}_0}\right) = 0 \ and \ \lim_{\widehat{N}_0 \to \infty} ln\left(\frac{j+\widehat{N}_0}{\widehat{N}_0}\right) = 0, \ then$$

L'Hospital's Rule may be used, so that

$$\lim_{\widehat{N}_0 \to \infty} \frac{ln\left(\dfrac{i+\widehat{N}_0}{\widehat{N}_0}\right)}{ln\left(\dfrac{j+\widehat{N}_0}{\widehat{N}_0}\right)} = \lim_{\widehat{N}_0 \to \infty} \frac{\dfrac{d}{d\widehat{N}_0} ln\left(\dfrac{i+\widehat{N}_0}{\widehat{N}_0}\right)}{\dfrac{d}{d\widehat{N}_0} ln\left(\dfrac{j+\widehat{N}_0}{\widehat{N}_0}\right)} =$$

$$\lim_{\widehat{N}_0 \to \infty} \frac{\dfrac{-i}{\widehat{N}_0 \cdot (i+\widehat{N}_0)}}{\dfrac{-j}{\widehat{N}_0 \cdot (j+\widehat{N}_0)}}. \ Applying \ the \ rule \ again,$$

$$\lim_{\widehat{N}_0 \to \infty} \frac{i \cdot \dfrac{d}{d\widehat{N}_0}(j+\widehat{N}_0)}{j \cdot \dfrac{d}{d\widehat{N}_0}(i+\widehat{N}_0)} = \frac{i}{j}.$$

Similarly, conditions for use of L'Hopital's Rule apply as $\widehat{N}_0 \to 0$ since

$$\lim_{\widehat{N}_0 \to 0} ln\left(\frac{i+\widehat{N}_0}{\widehat{N}_0}\right) = \infty \ and \ \lim_{\widehat{N}_0 \to 0} ln\left(\frac{j+\widehat{N}_0}{\widehat{N}_0}\right) = \infty.$$

$$So \ \lim_{\widehat{N}_0 \to 0} \frac{\dfrac{d}{d\widehat{N}_0} ln\left(\dfrac{i+\widehat{N}_0}{\widehat{N}_0}\right)}{\dfrac{d}{d\widehat{N}_0} ln\left(\dfrac{j+\widehat{N}_0}{\widehat{N}_0}\right)} = \lim_{\widehat{N}_0 \to 0} \frac{\dfrac{-i}{\widehat{N}_0 \cdot (i+\widehat{N}_0)}}{\dfrac{-j}{\widehat{N}_0 \cdot (j+\widehat{N}_0)}}$$

$$= \lim_{\widehat{N}_0 \to 0} \frac{i \cdot (j+\widehat{N}_0)}{j \cdot (i+\widehat{N}_0)} = \frac{i \cdot j}{j \cdot i} = 1.$$

For uniqueness of $\widehat{N}_0$, strictly increasing monotonicity must be shown to prove a one-to-one correspondence such that for any $a \in (0,\infty)$, $\delta \in (0,a)$, $\Delta \in (0,\infty)$, *and* $i \neq j$,

$$1 < \frac{\ln\left(\frac{i}{[a-\delta]}+1\right)}{\ln\left(\frac{j}{[a-\delta]}+1\right)} < \frac{\overline{s}\cdot\ln\left(\frac{i+a}{a}\right)}{\overline{s}\cdot\ln\left(\frac{j+a}{a}\right)} = \frac{\overline{s}_i}{\overline{s}_j} < \frac{\ln\left(\frac{i}{[a+\Delta]}+1\right)}{\ln\left(\frac{j}{[a+\Delta]}+1\right)} < \frac{i}{j},$$

for $i > j$.

In other words, assigning the functions $f, g : \mathbb{R}^+ \rightarrow \mathbb{R}^+$, it must be shown that

$$\frac{f\left(\widehat{N}_0\right)}{g\left(\widehat{N}_0\right)} = \frac{\ln\left(\frac{i+\widehat{N}_0}{\widehat{N}_0}\right)}{\ln\left(\frac{j+\widehat{N}_0}{\widehat{N}_0}\right)} \text{ is strictly increasing for all } \widehat{N}_0 \in (0,\infty)$$

and $i > j$.

To do this, first note that the functions f and g are both differentiable on the open interval $(0,\infty)$, and that,

$$f\left(\widehat{N}_0\right) = \lim_{\widehat{N}_0 \to \infty} \ln\left(\frac{i+\widehat{N}_0}{\widehat{N}_0}\right) = 0 \text{ and } \lim_{\widehat{N}_0 \to \infty} \ln\left(\frac{j+\widehat{N}_0}{\widehat{N}_0}\right) = 0.$$

Also note that,

$$g' = \frac{d}{d\widehat{N}_0} \ln\left(\frac{j+\widehat{N}_0}{\widehat{N}_0}\right) = \frac{-j}{\widehat{N}_0 \cdot (j+\widehat{N}_0)} < 0 \text{ for all } j > 0 \text{ and }$$

$\widehat{N}_0 \in (0,\infty).$

Under these conditions, and using (Pinelis, 2002, Proposition 1.1 (1)),

$$\text{if } \frac{f'}{g'} \text{ is increasing on} (0,\infty), \text{then} \left(\frac{f}{g}\right)' > 0 \text{ on} (0,\infty).$$

In other words, if $\left(\frac{f}{g}\right)' > 0$ over the open interval, then the ratio of the two functions f and g is strictly increasing over the open interval $(0,\infty)$. Since,

$$\frac{d}{d\widehat{N}_0}\left(\frac{f'}{g'}\right) = \frac{i \cdot j \cdot (i-j)}{\left(i \cdot j + j \cdot \widehat{N}_0\right)^2} > 0 \; for \; all \; i > j \; and \; \widehat{N}_0 \in (0,\infty).$$

then $\dfrac{f'}{g'}$ *is increasing on* $(0,\infty)$ *and* $\left(\dfrac{f}{g}\right)' > 0.$

Therefore $\dfrac{f\left(\widehat{N}_0\right)}{g\left(\widehat{N}_0\right)}$ *is strictly increasing as* $\widehat{N}_0$ *increases on the*

open interval $(0,\infty).$

Similarly, using (Pinelis, 2002, Proposition 1.1 (2)),

if $\dfrac{f'}{g'}$ *is decreasing on* $(0,\infty)$*, then* $\left(\dfrac{f}{g}\right)' < 0 \; on \; (0,\infty),$

it may be shown in the same manner that for any $\delta \in (0,a), \Delta \in (0,\infty), and \; i < j$, the ratio of the logarithms is strictly decreasing for all $a \in (0,\infty)$ such that

$$1 > \frac{\ln\left(\dfrac{i}{[a-\delta]}+1\right)}{\ln\left(\dfrac{j}{[a-\delta]}+1\right)} > \frac{\overline{s} \cdot \ln\left(\dfrac{i+a}{a}\right)}{\overline{s} \cdot \ln\left(\dfrac{j+a}{a}\right)} = \frac{\overline{s}_i}{\overline{s}_j} > \frac{\ln\left(\dfrac{i}{[a+\Delta]}+1\right)}{\ln\left(\dfrac{j}{[a+\Delta]}+1\right)} > \frac{i}{j},$$

for $i < j.$

Hence $\overline{s} \cdot ln\left(\dfrac{j+a}{a}\right) \cdot \dfrac{ln\left(\dfrac{i+\widehat{N}_0}{\widehat{N}_0}\right)}{ln\left(\dfrac{j+\widehat{N}_0}{\widehat{N}_0}\right)} = \overline{s} \cdot ln\left(\dfrac{i+a}{a}\right) = \overline{s}_i \; if \; and \; only \; if \; \widehat{N}_0 = a.$

∎

Corollary 2.3: For $i, j \in P$ *where*

$$P \in \{\text{elements representing simultaneous units of stimuli}\}$$

and $i \neq j$, for the same social situation, if

$$1 < \frac{\bar{s}_i}{\bar{s}_j} < \frac{i}{j} \ \text{if} \ i > j \ \text{or} \ 1 > \frac{\bar{s}_i}{\bar{s}_j} > \frac{i}{j} \ \text{if} \ i < j,$$

then for an expected value $E[S]$, there exists a unique solution for $\widehat{N}_0$ supporting Theorem 2.3.

Proof: This is derived directly from the proof of Theorem 2.3. ∎

Before proceeding, the concept of dissonance versus relevance and extent social deviation needs further discussion. When a group member observes a socially deviant event, that member will associate some relevance to it and some extent social deviation based on the immediate impact the event has on the member and extent in which the member identifies with the social norms of the group at the time. Individuals have different perceptions based on experience, but in what will be assumed a relatively cohesive group, all group members should experience some similarity of interpretation of relevance and extent social deviation within bounds. Let us assume for some specific social event e and given social situation that each member perceives the event's relevance R within the bounds $0 < a_1 \leq R_e \leq a_2 \leq 1$ and extent social deviation $0 < c_1 \leq E_e \leq c_2 \leq d$. Uniformly distributed intervals are assumed for $[a_1, a_2], [c_1, c_2] \subseteq (0, d]$. Defining $S_e = R_e \cdot E_e$ as the random variable representing dissonance for the particular social event and situation, then $Var(S_e) \leq Var(S)$.

Theorem 2.4:

Let $P = \{\text{observed units of simultaneous stimuli of the same type}\}$, and $n(O_p)$ be the number of observations for each condition $p \in P$. Assume all $\bar{s}_k$, $k \in P$, satisfy Corollary 2.3, and that the same social situation with geometric mean social noise $\widehat{N}_0$ is known for the dissonance-causing event having random variable S_e. Given $Var(S_e) \leq Var(S)$, then for some $\varepsilon > 0$,

$$\text{if} \ \bar{s}(\widehat{N}_0) = \frac{1}{|P|} \cdot \sum_{k \in P} \frac{\bar{s}_k}{ln\left(\frac{k + \widehat{N}_0}{\widehat{N}_0}\right)} \text{then} \ P\left[\left|\bar{s}(\widehat{N}_0) - E(S_e)\right| \geq \varepsilon\right]$$

$$\leq \frac{Var(S)}{\varepsilon^2 \cdot \sum_{k=1}^{max(p)} n(O_p)}.$$

where $|P|$ is the cardinality of P.

Proof: As per Equation 2.2, we have $\bar{s}_k = \bar{s} \cdot ln\left(\frac{k + \widehat{N}_0}{\widehat{N}_0}\right)$ where each $\bar{s}_k$ consists of multiple $n(O_k)$ observations of $s_k = S \cdot ln\left(\frac{k + \widehat{N}_0}{\widehat{N}_0}\right)$, so that

$$\overline{s}\left(\widehat{N}_0\right) = \frac{\overline{s}_k}{ln\left(\dfrac{k+\widehat{N}_0}{\widehat{N}_0}\right)} = \frac{\Sigma_{n(O_k)}\,s_k}{n(O_k)\cdot ln\left(\dfrac{k+\widehat{N}_0}{\widehat{N}_0}\right)}.$$

In general, for $|P| \geq 2,$

$$\overline{s}\left(\widehat{N}_0\right) = \frac{1}{|P|}\cdot \sum_{k\in P} \frac{\Sigma_{n(O_k)}\,s_k}{n(O_k)\cdot ln\left(\dfrac{k+\widehat{N}_0}{\widehat{N}_0}\right)} = \frac{1}{|P|}\cdot \sum_{k\in P} \frac{\overline{s}_k}{ln\left(\dfrac{k+\widehat{N}_0}{\widehat{N}_0}\right)}.$$

Therefore, using the weak law of large numbers and Chebyshev's Inequality (Pishro-Nik, 2014, pp. 378–379), and knowing that

$$Var\left(S\right) = \int_0^1 x^2\cdot ln\left(\frac{1}{x}\right)dx - \left[\int_0^1 x\cdot ln\left(\frac{1}{x}\right)dx\right]^2 = 0.111 - 0.25^2 = 0.049 \quad \text{is}$$

finite, for a given social situation and some form of deviation from expected social norms,

$$p\left[\overline{s}\left(\widehat{N}_0\right) - E[S_e]|\geq \varepsilon\right] \leq \frac{Var\left(S_e\right)}{\varepsilon^2\cdot \Sigma_{p\in P}\,n\left(O_p\right)} \leq \frac{Var\left(S\right)}{\varepsilon^2\cdot \Sigma_{p\in P}\,n\left(O_p\right)}. \qquad \blacksquare$$

We now have the necessary tools to estimate the value $\widehat{N}_0$.

Theorem 2.5: Define two sets

$P = \left\{\text{observed units of simultaneous stimuli of the same type}\right\}$ *and*

$$P_m = \left\{element\ pair\left(m,k\right) : \frac{\overline{s}_m}{\overline{s}_k}\ satisfying\ Corollary\ 2.3,\ k \neq m,\ and\ m,k \in P\right\}.$$

Assume for all $k,m \in P$, $n\left(O_k\right) \cong n\left(O_m\right)$. For the same social situation socially deviant event, based on Theorems 2.3 and 2.4, there exists a unique $\widehat{N}_0$, $\dfrac{1}{e-1} \leq \widehat{N}_0 < \infty$ such that if

$$\sum_{\substack{m\in P \\ |P_m|>0}} \left[\overline{s}_m - \frac{1}{|P_m|}\cdot \sum_{(m,k)\in P_m} \frac{\overline{s}_k\cdot ln\left(\dfrac{m+\widehat{N}_0}{\widehat{N}_0}\right)}{ln\left(\dfrac{k+\widehat{N}_0}{\widehat{N}_0}\right)}\right] = 0\ then\ \overline{s}\left(\widehat{N}_0\right) = \frac{1}{|P|}\cdot \sum_{k\in P} \frac{\overline{s}_k}{ln\left(\dfrac{k+\widehat{N}_0}{\widehat{N}_0}\right)}$$

is an unbiased estimator of $\bar{s}$ as a function of $\widehat{N}_0$.

Proof: Using Theorem 2.4, as $\sum_{p \in P} n(O_p) \to \infty$, $P\left[\left|E[S_e] - \bar{s}(\widehat{N}_0)\right| \geq \varepsilon\right] \to 0$
for any $\varepsilon > 0$ and finite $Var(S)$. Given

$$E\left[S_e \cdot ln\left(\frac{m + N_0}{N_0}\right)\right] = E[S_e] \cdot ln\left(\frac{m + \widehat{N}_0}{\widehat{N}_0}\right)$$

for some unique $\widehat{N}$, per the Strong Law of Large Numbers

$$\frac{\sum_{i=1}^{n(O_m)} S_{e,i}}{n(O_m)} = \frac{\bar{s}_m}{ln\left(\frac{m + \widehat{N}_0}{\widehat{N}_0}\right)} \to E[S_e] \text{ as } n(O_m) \to \infty.$$

Hence, based on Theorem 2.4,

$$\sum_{\substack{m \in P \\ |P_m| > 0}} \left[\bar{s}_m - \frac{1}{|P_m|} \cdot \sum_{(m,k) \in P_m} \frac{\bar{s}_k \cdot ln\left(\frac{m + \widehat{N}_0}{\widehat{N}_0}\right)}{ln\left(\frac{k + \widehat{N}_0}{\widehat{N}_0}\right)}\right] \to 0$$

$$as \ n(O_m) \to \infty, and \ |P| > |P_m|.$$

Therefore, for some $\widehat{N}_0$, such that

$$\sum_{\substack{m \in P \\ |P_m| > 0}} \left[\bar{s}_m - \frac{1}{|P_m|} \cdot \sum_{(m,k) \in P_m} \frac{\bar{s}_k \cdot ln\left(\frac{m + \widehat{N}_0}{\widehat{N}_0}\right)}{ln\left(\frac{k + \widehat{N}_0}{\widehat{N}_0}\right)}\right] = 0,$$

$$then \ \bar{s}(\widehat{N}_0) = \frac{1}{|P|} \cdot \sum_{k \in P} \frac{\bar{s}_k}{ln\left(\frac{k + \widehat{N}_0}{\widehat{N}_0}\right)}$$

is an unbiased estimator for the sample mean $\bar{s}(\widehat{N}_0)$. ∎

Example application: Using Table 2.3 values for $\bar{s}_k$ from the crowd-gathering experiment, we solve for $\bar{s}(\widehat{N}_0)$ using Theorem 2.4 by varying $\widehat{N}_0$ while applying Theorem 2.5.
The first step is to define the sets P and P_m such that

$P = \{1, 2, 3, 5, 10, 15\}.$

$P_1 = \{(1,2),(1,3),(1,5),(1,10),(1,15)\},$

$P_2 = \{(2,1),(2,3),(2,5),(2,10),(2,15)\},$

$P_3 = \{(3,1),(3,2),(3,5),(3,10),(3,15)\},$

$P_5 = \{(5,1),(5,2),(5,3),(5,15)\} : (5,10) \notin P_5 \ since \ \dfrac{\overline{s}_5}{\overline{s}_{10}}$

$= \dfrac{0.422}{0.394} > 1,$

$P_{10} = \{(10,1),(10,2),(10,3),(10,15)\} : (10,5) \notin P_{10} \ since \ \dfrac{\overline{s}_{10}}{\overline{s}_5}$

$= \dfrac{0.394}{0422} < 1, and$

$P_{15} = \{(15,1),(15,2),(15,3),(15,5),(15,10)\}.$

To apply Theorem 2.5, note from Figure 2.1 that $n(O_1) = 232$, $n(O_2) = 212$, $n(O_3) = 211$, $n(O_5) = 222$, $n(O_{10}) = 268$, *and* $n(O_{15}) = 279$ are all approximately equal.

For some real number value $\dfrac{1}{e-1} \leq \widehat{N}_0 < \infty$, with 1 being a good starting point, iterate the value for $\widehat{N}_0$ into the summation equation until the sum equals as close to zero as you want to go (using a spreadsheet makes this easier). As per Theorem 2.5, to find $\widehat{N}_0$ solve for

$$\sum_{\substack{m \in P \\ |P_m| > 0}} \left[\overline{s}_m - \frac{1}{|P_m|} \cdot ln\left(\frac{m + \widehat{N}_0}{\widehat{N}_0}\right) \cdot \sum_{(m,k) \in P_m} \frac{\overline{s}_k}{ln\left(\dfrac{k + \widehat{N}_0}{\widehat{N}_0}\right)} \right] = 0.$$

Summing over all values of $m \in P$ and P_m *such that* $|P_m| > 0$ for the crowd-gathering experiment,

$$0.145 - \frac{1}{5} \cdot ln\left(\frac{1+\widehat{N}_0}{\widehat{N}_0}\right) \cdot \left[\frac{0.241}{ln\left(\frac{2+\widehat{N}_0}{\widehat{N}_0}\right)} + \frac{0.278}{ln\left(\frac{3+\widehat{N}_0}{\widehat{N}_0}\right)} + \frac{0.422}{ln\left(\frac{5+\widehat{N}_0}{\widehat{N}_0}\right)} \right.$$

$$\left. + \frac{0.394}{ln\left(\frac{10+\widehat{N}_0}{\widehat{N}_0}\right)} + \frac{0.515}{ln\left(\frac{15+\widehat{N}_0}{\widehat{N}_0}\right)}\right]$$

$$+ 0.241 - \frac{1}{5} \cdot ln\left(\frac{2+\widehat{N}_0}{\widehat{N}_0}\right)$$

$$\cdot \left[\frac{0.145}{ln\left(\frac{1+\widehat{N}_0}{\widehat{N}_0}\right)} + \frac{0.278}{ln\left(\frac{3+\widehat{N}_0}{\widehat{N}_0}\right)} + \frac{0.422}{ln\left(\frac{5+\widehat{N}_0}{\widehat{N}_0}\right)} + \frac{0.394}{ln\left(\frac{10+\widehat{N}_0}{\widehat{N}_0}\right)} \right.$$

$$\left. + \frac{0.515}{ln\left(\frac{15+\widehat{N}_0}{\widehat{N}_0}\right)}\right]$$

$$+ 0.278 - \frac{1}{5} \cdot ln\left(\frac{3+\widehat{N}_0}{\widehat{N}_0}\right)$$

$$\cdot \left[\frac{0.145}{ln\left(\frac{1+\widehat{N}_0}{\widehat{N}_0}\right)} + \frac{0.241}{ln\left(\frac{2+\widehat{N}_0}{\widehat{N}_0}\right)} + \frac{0.422}{ln\left(\frac{5+\widehat{N}_0}{\widehat{N}_0}\right)} + \frac{0.394}{ln\left(\frac{10+\widehat{N}_0}{\widehat{N}_0}\right)} \right.$$

$$\left. + \frac{0.515}{ln\left(\frac{15+\widehat{N}_0}{\widehat{N}_0}\right)}\right]$$

$$+\,0.422 - \frac{1}{4}\cdot\left(\frac{5+\widehat{N}_0}{\widehat{N}_0}\right)$$

$$\cdot\left[\frac{0.145}{ln\left(\dfrac{1+\widehat{N}_0}{\widehat{N}_0}\right)}+\frac{0.241}{ln\left(\dfrac{2+\widehat{N}_0}{\widehat{N}_0}\right)}+\frac{0.278}{ln\left(\dfrac{3+\widehat{N}_0}{\widehat{N}_0}\right)}+\frac{0.515}{ln\left(\dfrac{15+\widehat{N}_0}{\widehat{N}_0}\right)}\right]$$

$$+\,0.394 - \frac{1}{4}\cdot ln\left(\frac{10+\widehat{N}_0}{\widehat{N}_0}\right)$$

$$\cdot\left[\frac{0.145}{ln\left(\dfrac{1+\widehat{N}_0}{\widehat{N}_0}\right)}+\frac{0.241}{ln\left(\dfrac{2+\widehat{N}_0}{\widehat{N}_0}\right)}+\frac{0.278}{ln\left(\dfrac{3+\widehat{N}_0}{\widehat{N}_0}\right)}+\frac{0.515}{ln\left(\dfrac{15+\widehat{N}_0}{\widehat{N}_0}\right)}\right]$$

$$+\,0.515 - \frac{1}{5}\cdot ln\left(\frac{15+\widehat{N}_0}{\widehat{N}_0}\right)$$

$$\cdot\left[\frac{0.145}{ln\left(\dfrac{1+\widehat{N}_0}{\widehat{N}_0}\right)}+\frac{0.241}{ln\left(\dfrac{2+\widehat{N}_0}{\widehat{N}_0}\right)}+\frac{0.278}{ln\left(\dfrac{3+\widehat{N}_0}{\widehat{N}_0}\right)}+\frac{0.422}{ln\left(\dfrac{5+\widehat{N}_0}{\widehat{N}_0}\right)}+\frac{0.394}{ln\left(\dfrac{10+\widehat{N}_0}{\widehat{N}_0}\right)}\right]\cdot$$

Summing all six calculations using the value $\widehat{N}_0 = 0.6299$ results in $-0.0109 + 0.0113 - 0.0070 + 0.0714 - 0.0583 - 0.0065 = 0.0$.

To find $\overline{s}\left(\widehat{N}_0\right)=\overline{s}$, using Theorem 2.5,

$$\overline{s}\left(0.6299\right)=\left[\frac{0.145}{ln\left(\dfrac{1+\widehat{N}_0}{\widehat{N}_0}\right)}+\frac{0.241}{ln\left(\dfrac{2+\widehat{N}_0}{\widehat{N}_0}\right)}+\frac{0.278}{ln\left(\dfrac{3+\widehat{N}_0}{\widehat{N}_0}\right)}+\frac{0.422}{ln\left(\dfrac{5+\widehat{N}_0}{\widehat{N}_0}\right)}\right.$$

$$\left.+\frac{0.394}{ln\left(\dfrac{10+\widehat{N}_0}{\widehat{N}_0}\right)}+\frac{0.515}{ln\left(\dfrac{15+\widehat{N}_0}{\widehat{N}_0}\right)}\right]\cdot\frac{1}{6}=0.162.$$

This ends the example application of Corollary 2.3 and Theorems 2.3, 2.4, and 2.5.

From Equation 2.2, using $\bar{s} = 0.162$ and $\hat{N}_0 = 0.63$, Figure 2.6 displays the 95 percent confidence intervals about the empirical data, Latane's (1981) psychosocial power law as the dashed line, and the theoretical values in probability space using $F_S\left(\bar{s}_p\right)$ with $d = 1$ as the solid line.

Calculated values for $\bar{s}_p$ and $F_S\left(\bar{s}_p\right)$ along with the associated 95 percent confidence intervals are provided in Table 2.4.

Based on the derived example values, using $\bar{s}_p = 0.162 \cdot ln\left(\dfrac{p + 0.63}{0.63}\right)$ and transforming it to probability space result in a mean square error value of 0.00126 for

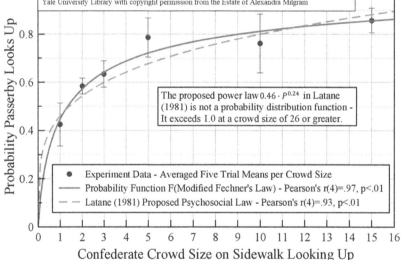

Milgram's 1969 Crowd Gathering Experiment
(Experiment Data from Stanley Milgram Papers, Yale University Archives, Group # 1406)

FIGURE 2.6 Depicting Equation 2.2 Where $\bar{s} = 0.162$, $\hat{N}_0 = 0.63$, and d = 1, Latane's Function Compared to Milgram's Empirical Mean Probability of Looking Up Values by Crowd Size within 95 Percent Confidence Intervals.

TABLE 2.4 Confidence Intervals Using Milgram's Empirical Data Compared to Calculated Results

Crowd Condition	p = 1	p = 2	p = 3	p = 5	p = 10	p = 15
95% Confidence Interval	(.34, .51)	(.55, .62)	(.58, .69)	(.71, .87)	(.64, .88)	(.81, .91)
Empirical Probability (Stanley Milgram Papers, MS 1406)	0.425	0.584	0.634	0.786	0.761	0.857
$\bar{s}_p = \bar{s} \cdot ln\left(\dfrac{p + \hat{N}_0}{\hat{N}_0}\right) =$	0.154	0.231	0.283	0.354	0.457	0.519
$For\, d = 1, F_S\left(\bar{s}_p\right) =$	0.442	0.570	0.640	0.722	0.815	0.859

the six conditions. With that, and all values falling within the 95 percent confidence bounds, the operational efficacy of Theorem 2.5, using independent empirical data for this social situation, has been shown to be adequate under uniform encoding.

2.3.2 Encoding: Exponential Distribution

As noted at the beginning of Section 2.2, the problem with Equation 2.1 is that as p gets larger and tends toward infinity (i.e., $p \to \infty$), so does s_p such that for some $N_0 \geq \dfrac{1}{e-1}$

$$\lim_{p \to \infty} s_P = S \cdot ln \frac{p+N_0}{N_0} \to \infty.$$

Clearly, this is not a bounded case leading to a logarithmic solution via the Cauchy Equations. Neither is this supported under optimal uniform density encoding based on information theory. If all assumptions and proposed theory to this point are correct, it would seem that an alternate functional approach is needed. Based on the four Cauchy functional equations and their associated criteria, the alternative must involve the exponential distribution with sensation magnitude es_p. Alternatively, to maximize Shannon's Measure of Information, and to maintain simplicity, we must have (Ben-Naim, 2017, pp. 105–106)

$$\int_0^\infty g\left(es_p\right) d\left(es_p\right) = 1 \text{ and } \int_0^\infty es_p \cdot g\left(es_p\right) d\left(es_p\right) = 1.$$

Probability of reaction is then represented by the general exponential cumulative distribution function (CDF):

$$\int_0^{es_p} g\left(es_p\right) d\left(es_p\right) = P\left(ES \leq es_p\right) = G_{ES}\left(es_p\right)$$

$$= 1 - e^{-es \cdot ln \frac{p+N_0}{N_0}} = 1 - e^{-es_p}. \qquad \text{\textit{Equation 2.3}}$$

TABLE 2.5 Empirical $\overline{es}_p$ Values Derived Using the Inverse Exponential Distribution; Empirical Data Derived from the Stanley Milgram Papers (MS 1406)

Crowd Condition	$p = 1$	$p = 2$	$p = 3$	$p = 5$	$p = 10$	$p = 15$
Empirical Probability (A_p)	0.425	0.584	0.634	0.786	0.761	0.857
$\overline{es}_p = -ln\left[1 - A_p\right]$	0.553	0.877	1.005	1.542	1.431	1.945

Using the same approach as was used in Section 2.3.1, we can derive $\overline{es}_p$, the sample mean of es_p, by manipulating Equation 2.3 to obtain the inverse exponential distribution

$$\overline{es}_p = G_{ES}^{-1}\left[G_{ES}\left(\overline{es}_p\right)\right] = -ln\left[1 - G_{ES}\left(\overline{es}_p\right)\right].$$

Using the $\overline{es}_p$ values from Table 2.5 and designating es as the unit stimulus sensation magnitude under exponential encoding when $N_0 = \dfrac{1}{e-1}$, we should find that

$$\overline{es}_1 = \overline{es}\cdot\left(\frac{1+\widehat{N}_0}{\widehat{N}_0}\right) \simeq 0.553,$$

$$\overline{es}_2 = \overline{es}\cdot ln\left(\frac{2+\widehat{N}_0}{\widehat{N}_0}\right) \simeq 0.877,\ and\ so\ on.$$

As in Section 2.3.1, based on Equation 2.2, this leads to showing that

$$\overline{es}_p = \overline{es}\cdot ln\left(\frac{p+\widehat{N}_0}{\widehat{N}_0}\right) for\ p \in P\ as\ defined\ in\ Theorem\ 2.5.$$

To calculate $\overline{es}$, Theorem 2.5 is again used, but instead of $\overline{s}_p$ under uniform encoding, the exponentially based $\overline{es}_p$ values from Table 2.5 are used. For observed stimuli condition set $P = \{1,2,3,5,10,15\}$, we must solve for $\widehat{N}_0$ such that

$$\sum_{\substack{m\in P \\ |P_m|>0}}\left[\overline{es}_m - \frac{1}{|P_m|}\cdot ln\left(\frac{m+\widehat{N}_0}{\widehat{N}_0}\right)\cdot \sum_{(m,k)\in P_m}\frac{\overline{es}_k}{ln\left(\frac{k+\widehat{N}_0}{\widehat{N}_0}\right)}\right] = 0$$

This results in $\widehat{N}_0 = 0.6095$ and $\overline{es} = \overline{es}\left(0.6095\right) = .588.$ Using

$$\overline{es}_p = 0.588\cdot ln\left(\frac{p+0.6095}{0.6095}\right)$$

and transforming it into probability space results in a mean square error value of 0.0011 for the six conditions when compared to empirical data from Milgram's crowd-gathering experiment. Table 2.6 demonstrates that both the cumulative distribution functions $F_S\left(\overline{s}_p\right)$ and $G_{ES}\left(\overline{es}_p\right)$ are exceptionally close to one another and

TABLE 2.6 Summary Table of Findings and Comparison with Empirical Data. Empirical Data Derived from the Stanley Milgram Papers (MS 1406)

Crowd Condition	$p = 1$	$p = 2$	$p = 3$	$p = 5$	$p = 10$	$p = 15$
95% Confidence Interval	(.34, .51)	(.55, .62)	(.58, .69)	(.71, .87)	(.64, .88)	(.81, .91)
Empirical Probability	0.425	0.584	0.634	0.786	0.761	0.857
Theoretical Uniform $F_S(\bar{s}_p)$	0.443	0.571	0.641	0.722	0.815	0.860
Theoretical Exponent $G_{ES}(\bar{es}_p)$	0.435	0.575	0.649	0.729	0.814	0.852
$\bar{es}_p = -ln\left[1 - G_{ES}(\bar{es}_p)\right]$	0.571	0.856	1.046	1.306	1.681	1.910

to the empirical results. Using difference equations, proving one or the other as the correct CDF with this data is not possible based on the large credibility intervals. The importance of using $G_{ES}(\bar{es}_p)$ is apparent though as $\bar{s}_p, \bar{es}_p \to \infty$.

Since $F_S(s_p)$ is bounded, there exists some finite value s_p, say $s_p = U$, which for any given situation and number of stimuli present, $\lim_{s_p \to U} F_S(s_p) = 1.0$. In the case of the unbounded es_p, $\lim_{es_p \to \infty} G_{ES}(es_p) = 1.0$. The implication under $F_S(s_p)$ is that after reaching a reaction probability of 1, which is difficult to comprehend in some situations, a decreasing probability of reaction with increasing stimulus results, eventually becoming negative. With a choice before us, one that is shown to be so close as to be unverifiable without a significant amount of data, the exponential encoding approach is chosen. This allows flexibility by not requiring a bounded sensation magnitude, but, more importantly, it is much easier to work with mathematically, and it leads to an equivalence relation with the power law. It is my opinion, based on data, that both play a role, but that is for others who are more capable to confirm or justifiably refute. It may be from an evolutionary standpoint that bounded sensation magnitude served its purpose well with limited stimuli, but as human group sizes increased, an unbounded but equally efficient information processing approach was needed.

As has already been assumed, let us now clarify in more detail the need for a noise floor. A noise floor is defined here as the lowest possible social noise intensity N_0. Given Equation 2.2, a value of $N_0 = 0$ is not defined, so there must be some value of $N_0 > 0$ at which the noise intensity cannot go below. Theoretically, if there is no social noise or inhibitions when observing a unit-stimuli, we should sense that

$$es \cdot ln\frac{1 + N_e}{N_e} = es, or \, that \, ln\frac{1 + N_e}{N_e} = 1$$

$$ln\frac{1 + N_0}{N_0} = 1 \, implies \, N_0 = \frac{1}{e - 1}$$

$$= 0.582 \, so \, that \, 1 - e^{-es \cdot ln\frac{1 + 0.582}{0.582}} = 1 - e^{-es}.$$

If we can assume $\overline{s}_1$ and $\overline{es}_1$ are bounded above by $\overline{s}$ and $\overline{es}$, respectively, then $N_e = \dfrac{1}{e-1} = 0.582$ should be the absolute lower bound for noise intensity. Such a noise level would indicate certainty (possibly complete understanding) of the socially deviant behavior with no inhibitions for a member's overt reaction. Now, the next question to answer is whether noise values change under encoding translation. This leads to an interesting relation that will be useful when comparing results between exponential encoding to uniform encoding.

From

$$\Delta s_{p,n} = s \cdot ln\left(\frac{p+N_0}{N_0}\right) - s \cdot ln\left(\frac{n+N_0}{N_0}\right) = s \cdot ln\left(\frac{p+N_0}{n+N_0}\right), n \le p$$

we observe that the constant unit stimuli sensation magnitude s for the situation and person may be divided out or multiplied by another constant, but that the values within the natural logarithmic functions must remain for the relationship with the data to be true. Given this observation, we may now reconsider our use of $d = 1$ in Theorem 2.2.

Assume that relevance r is scaled as before so that $r \in (0,1)$. If we make the big assumption – which may be correct and is certainly more interesting (and easier) – that $R \cdot E = S = ES$, then using Theorem 2.2 and previous results for $\overline{s}$ and $\overline{es}$, and for $e \in R_E$,

$$e \in \left(0, \frac{\overline{es}}{\overline{s}}\right] = \left(0, \frac{0.5882}{0.1618}\right] = (0, 3.6354], \text{ so that } d = 3.6354.$$

If we want to minimize the overall error between the logarithmic and exponential density functions, then $e \in (0, 3.79]$. As a final and very interesting alternative, instead of $e \in (0,1]$ as we initially started with, we use $e \in (0, 3.76174]$ so that $d = 3.76174$, then,

$$\frac{1}{3.76174} \cdot \left[1 - ln\left(\frac{1}{3.76174}\right)\right] = 0.61803, \text{ or the inverse of}$$

the golden ratio.

Given how many times the inverse of the golden ratio comes up here and in other psychological experiments (Gross and Miller, 1997; Benjafield and Adams-Weber, 1976), the last is worth further analytic and investigative effort for the future. For now, to minimize average error between the two densities, the interval used for e will be $(0, 3.79]$. Using this interval, for any sensation magnitude value $\overline{s}_p$ under logarithmic encoding, the density and cumulative density functions are

$$f_S\left(\bar{s}_p\right) = \frac{1}{3.79} \cdot ln\frac{3.79}{\bar{s}_p} \text{ and } F_S\left(\bar{s}_p\right) = \frac{\bar{s}_p}{3.79} \cdot \left[1 - ln\left(\frac{\bar{s}_p}{3.79}\right)\right]. \quad \textit{Equation 2.4}$$

Until more data is available, or a relationship with any existing proven theory can be made, Equation 2.4 should be considered an approximation when translating from uniform to exponential encoding and vice versa. In the end, the most accurate transformation from $\overline{es}$ to $\bar{s}$ is what is of interest since this provides access when working with exponential encoding to the sample mean unit stimulus sensation magnitude $\bar{s} = \bar{S}_e$ (relevance random variable multiplied by expected social extent deviation random variable within the given situation and event) and thereby allows a standard approach for comparing stimuli of various types in various social situations for people of various cultures. We may also use values obtained for $\overline{es}$ to compare with Stevens-measured exponent power law values derived from physical stimuli, such as the smell of heptane with a measured exponent of 0.6 (Stevens, 1975, p. 15). In this case, the exponents for heptane (0.6) and the crowd-gathering experiment (0.589) are similar in their sensation magnitude.

2.4 Fechner's Law and the Power Law

It has already been shown that Latane's (1981) psychosocial power law is not a CDF as was needed. Whether it is appropriate as a density function for the crowd-gathering data is now considered. Stevens (1975, pp. 13–14) – using his nomenclature – indicates that by taking the natural logarithm of

$$\psi = k \cdot \phi^\beta, \text{ we obtain}$$

$$ln(\psi) = ln(k \cdot \phi^\beta) = ln(k) + \beta \cdot ln(\phi) \text{ for stimulus magnitude } \phi.$$

Setting $ln\left(\psi_p\right) = es_p, k = 1, \beta = \overline{es}$, and $ln(\phi) = ln\frac{p + \widehat{N}_0}{\widehat{N}_0}$, the equivalent power law function for the crowd-gathering experiment in power space may be stated as

$$\psi_p = \left(\frac{p + \widehat{N}_0}{\widehat{N}_0}\right)^{\overline{es}}. \quad \textit{Equation 2.5}$$

Through further equivalence modification of Equation 2.5, we obtain

$$\psi_p = \left(\frac{p + \widehat{N}_0}{\widehat{N}_0}\right)^{\overline{es}} = \left(\frac{1}{\widehat{N}_0}\right)^{\overline{es}} \cdot \left(p + \widehat{N}_0\right)^{\overline{es}}.$$

This form of Stevens power law is of the same form as derived by Zwislocki (1965, Eq. 204, p. 84). The cumulative probability distribution function $G_{ES}\left[ln(\psi)\right]$ may be used with Stevens power law by simply noting that

$$G_{ES}\left[ln(\psi_p)\right]=1-e^{-ln(\psi_p)}=1-e^{-es_p}=1-\left(\frac{\widehat{N}_0}{p+\widehat{N}_0}\right)^{\overline{es}}=1-\frac{1}{\psi_p}$$

Equation 2.6

It is important to recognize that the inverse relation predicted by Sherif and Sherif (1956, pp. 170–171), Zipf (1948, Loc 10890), and Festinger (1957, p. 18) is again present. As stimulus intensity increases, dissonance sensation magnitude increases, and the pressure to reduce the dissonance (a.k.a. probability of reaction) increases. As a final check, using the crowd-gathering result with two confederates who are looking up, comparing the power law density function of Equation 2.5 with Equation 2.3, as expected we get

$$\psi_2=\left(\frac{2+0.6095}{0.6095}\right)^{0.589}=2.354,\ so$$

$$G_{ES}\left[ln(2.354)\right]=1-e^{-ln(2.354)}=1-\left(\frac{0.6095}{2+0.6095}\right)^{0.589}=0.575.$$

As should be expected, this is the same result derived using $\overline{es}$ based on exponential encoding and Fechner's Law as shown in Table 2.6. As used here, and as argued independently by Shepard (1981, pp. 35–48), it would seem that the power law is merely a transform of Fechner's Law under exponential encoding in probability space for the applications as presented, which could explain Stevens' (1975, p. 13) observation that the exponent β seems to be a function of the type of stimulus. Finally, we may conclude the power law-like approach that Latane (1981) used was for conceptual purposes only, and for that along with his other significant achievements he is greatly appreciated and respected by this author. It is hoped this work builds on and adds something to his Social Impact Theory, as it is certainly a result of it.

2.5 Summary

This finishes the development of the basic mathematical structure. Having built on the work of Bibb Latane, the next step is a quantitative analysis of simple social systems through further development which builds on this basic structure. For this development to occur, we explore data from two experiments, both involving the western first come first-served queue. The first experiment for analysis was performed by Schmitt et al. (1992). This experiment provides the numerical key that unlocks data from the second and more socially complex 1978 field experiment performed by Stanley Milgram's graduate students and documented in Milgram et al. (1986).

References

Arons, L., & Irwin, F. (1932). Equal Weights and Psychophysical Judgements. *Journal of Experimental Psychology, 15*(6), 733–751. https://doi.org/10.1037/h0070521

Benjafield, J., & Adams-Weber, J. (1976). The Golden Section Hypothesis. *British Journal of Psychology, 67*(1), 11–15. https://doi.org/10.1111/j.2044-8295.1976.tb01492.x

Ben-Naim, A. (2017). *Information Theory–Part 1: An Introduction to the Fundamental Concepts.* World Scientific Publishing Co. https://doi.org/10.1142/10417

Devore, J. (1987). *Probability and Statistics for Engineering and the Sciences* (2nd ed.). Brookes/Cole Publishing, a Division of Wadsworth, Inc.

Dio, C. (1917). *Dio Cassius Roman History, Books 51–55* (Jeffrey Henderson, Ed., Earnest Cary, Trans.). Harvard University Press.

Fernberger, S. (1913). On the Relation of the Methods of Just Perceptible Differences and Constant Stimuli. *The Psychological Monographs, 14*(4), 1–81. https://doi.org/10.1037/h0093068

Festinger, L. (1957). *A Theory of Cognitive Dissonance.* Stanford University Press. Retrieved from www.sup.org/books/title/?id=3850

Fiske, S., & Taylor, S. (2017). *Social Cognition–From Brains to Culture* (3rd ed.). Sage Publications.

Glen, A., Leemis, L., & Drew, J. (2004). Computing the Distribution of the Product of Two Continuous Random Variables. *Computational Statistics and Data Analysis, 44*(3), 451–464. https://doi.org/10.1016/S0167-9473(02)00234-7

Green, D., & Swets, J. (1988). *Signal Detection Theory and Psychophysics.* Peninsula Publishing.

Gross, S., & Miller, N. (1997). The "Golden Section" and Bias in Perceptions of Social Consensus. *Personality and Social Psychology Review, 1*(3), 241–271. https://doi.org/10.1207/s15327957pspr0103_4

Guilford, J. (1954). *Psychometric Methods* (2nd ed.). McGraw-Hill Book Co.

Hautus, M., Macmillan, N., & Creelman, C. (2022). *Detection Theory: A User's Guide* (3rd ed.). Routledge Publishers. https://doi.org/10.4324/9781003203636

Hellstromg, A., & Rammsayer, T. (2015). Time-Order Errors and Standard-Position Effects in Duration Discrimination: An Experimental Study and an Analysis By the Sensation-Weighting Model. *Attention, Perception, & Psychophysics, 77*, 2409–2423. https://doi.org/10.3758/s13414-015-0946-x

Holman, E.W., & Marley, A.A. (1974). Stimulus and Response Measurement. In E.C. Carterette & M.P. Friedman (Eds.), *Handbook of Perception, Vol II–Psychophysical Judgement and Measurement* (pp. 173–213). Academic Press. https://doi.org/10.1016/B978-0-12-161902-2.X5001-9

Latane, B. (1981). The Psychology of Social Impact. *American Psychologist, 36*(4), 343–356. https://doi.org/10.1037/0003-066X.36.4.343

Laughlin, S. (1981). A Simple Coding Procedure Enhances a Neuron's Information Capacity. *Zeitschrift für Naturforschung, 36*, 910–912. https://doi.org/10.1515/znc-1981-9-1040

Milgram, S. (1970). The Experience of Living in Cities. *Science, 167*(3924), 1461–1468. https://doi.org/10.1126/science.167.3924.1461

Milgram, S., Bickman, L., & Berkowitz, L. (1969). Note on the Drawing Power of Crowds of Different Size. *Journal of Personality and Social Psychology, 13*(2), 79–82. https://doi.org/10.1037/h0028070

Milgram, S., Liberty, J., Toledo, R., & Wackenhut, J. (1986). Response to Intrusion into Waiting Lines. *Journal of Personality and Social Psychology, 51*(4), 683–689. https://doi.org/10.1037/0022-3514.51.4.683

Pham-Gia, T., & Turkkan, N. (2002). The Product and Quotient of General Beta Distributions. *Statistical Papers, 43*, 537–550. https://doi.org/10.1007/S00362-002-0122-Y

Pinelis, I. (2002). L'Hospital Type Rules for Monotonicity, with Applications. *Journal of Inequalities in Pure and Applied Mathematics, 3*(1), Article 5, 1–5. Retrieved from www.emis.de/journals/JIPAM/images/157_05_JIPAM/157_05.pdf

Pishro-Nik, H. (2014). *Introduction to Probability, Statistics, and Random Processes.* Kappa Research LLC. Retrieved from https://scholarworks.umass.edu/ece_ed_materials/1/

Pratt, C. (1933). The Time-Error in Psychological Judgements. *The American Journal of Psychology, 45*(2), 292–297. https://doi.org/10.2307/1414280

Rohatgi, V. (1976). *An Introduction to Probability Theory Mathematical Statistics*. John Wiley & Sons. https://doi.org/10.1002/bimj.4710210813

Schmitt, B., Dube, L., & Leclerc, F. (1992). Intrusions Into Waiting Lines: Does the Queue Constitute a Social System? *Journal of Personality and Social Psychology, 63*(5), 806–815. https://doi.org/10.1037/0022-3514.63.5.806

Shepard, R. (1981). Psychological Relations and the Psychophysical Scales: On the Status of "Direct" Psychophysical Measurement. *Journal of Mathematical Psychology, 24*, 21–57. https://doi.org/10.1016/0022-2496(81)90034-1

Sherif, M., & Sherif, C. (1956). *An Outline of Social Psychology*. Harper & Brothers.

Steblay, N.M. (1987). Helping Behavior in Rural and Urban Environments: A Meta-Analysis. *Psychological Bulletin, 102*(3), 346–356. https://doi.org/10.1037/0033-2909.102.3.346

Stevens, S.S. (1975). *Psychophysics*. John Wiley and Sons Inc.

Stone, J. (2015). *Information Theory: A Tutorial Introduction*. Sebtel Press (Kindle Edition). https://doi.org/10.48550/arXiv.1802.05968

Tanner, T., Haller, R., & Atkinson, R. (1967). Signal Recognition as Influenced By Presentation Schedules. *Perception and Psychophysics, 2*(8), 349–358. https://doi.org/10.3758/BF03210070

Zipf, G. (1948). *Human Behavior and the Principle of Least Effort: An Introduction to Human Ecology*. Ravenio Books (Kindle Edition). https://doi.org/10.1002/1097-4679(195007)6:3<306::AID-JCLP2270060331>3.0.CO;2-7

Zwislocki, J.J. (1965). An Analysis of Some Auditory Characteristics. In R.D. Luce, R.R. Bush, & E. Galanter (Eds.), *Handbook of Mathematical Psychology (Vol. III)*. John Wiley & Sons. https://doi.org/10.1177/001316446802800339

3

REVISITING MILGRAM'S 1978 "RESPONSE TO INTRUSION INTO WAITING LINES" EXPERIMENT

This analysis shows that cost alone cannot account for all our results; there is an underlying structure to the situation, linked to the linear spatial configuration of the queue.

Milgram et al. (1986, p. 688), with permission
from the American Psychological Association

Dr. Stanley Milgram was prodigious at creating or facilitating the design and implementation of what are nothing less than brilliantly devised experiments. One of those experiments, conducted by his graduate students in the spring of 1978 and published shortly after his death (Milgram et al., 1986), relates to how queue members in typical New York locations react to a simultaneous intrusion by one or two student confederate members. Experimental trials were performed 20 or more times for each of six distinct conditions (zero, one, or two non-reacting buffers, and one or two intruders) using 17 locations in New York City where queues regularly formed. Data contained in Milgram et al. (1986) along with summary data and three of the original 1978 graduate student reports from the Stanley Milgram Papers (MS 1406) allow further analysis and model development to continue from the previous chapter. Results of this effort, using exponential encoding, lead to a significantly expanded and coherent interpretation of social interaction and structure within the linear spatial configuration of the queue as a social system.

3.1 Establishing an Analytic Model of Intragroup Social Impact

Six years after the article by Milgram et al. (1986) is published, Schmitt et al. (1992) publish an equally important paper further exploring the queue as a social system. The importance of this paper is that it provides a critical complement to the 1986 paper by Stanley Milgram et al. The paper by Schmitt et al. involves four studies – three were attitudinal questionnaires, and the fourth was a field experiment conducted at New York's Grand Central Station. It is the fourth study – the field experiment – that will be made use of.

DOI: 10.4324/9781003325161-4

In the fourth study, Schmitt et al. (1992) collect data consisting of 120 observations (four conditions with 30 observations each). Schmitt and his team use three control factors in this field experiment to focus specifically on the individual queue member who is intruded in front of by a student confederate. The control factors, designed to minimize influence from queue members near the member of interest, are:

1. All observations are from a single social situation – Grand Central railroad station ticket counter, New York City.
2. There is always a confederate in queue ahead of the intrusion point to eliminate reaction from members ahead of the intruder.
3. In the high social obligation case, two people in queue behind the member of interest are confederates who join the queue right after the member of interest joins.

In addition to the three control factors, Schmitt et al. (1992) consider four intrusion scenarios (i.e., conditions). The four scenarios in their field experiment are:

1. A <u>confederate intruding in front of the last queue member in queue</u> (illegitimate/ low social obligation).
2. A <u>confederate intruding in front of the third from the last member in queue</u> (illegitimate/high social obligation).
3. A <u>confederate joining the queue in front of the last queue member in queue</u>, while <u>acting as a friend</u> to the student confederate who is in line ahead of the queue member (legitimate/low social obligation).
4. A <u>confederate joining the queue in front of the third-to-last queue member in queue</u>, while <u>acting as a friend</u> to the student confederate who is in line ahead of the queue member (legitimate/high social obligation).

Reaction by the queue member of interest, directed at the intruder, falls into one of three reaction categories directed at the intruder. The reaction categories, used also by Milgram et al. (1986), are indirect objection (i.e., body language or indirect comments), direct verbal objection, or physical action used to gain the attention of and eject the intruder. Probability of reaction for each condition is shown in Table 3.1.

Findings from this and the Milgram intrusion data to be presented is that the greater the sensation magnitude representing the group member of interest's dissonance, the

TABLE 3.1 Probability of a Reaction, Given the Experimental Condition (Schmitt et al., 1992, p. 813 Table 1) with Permission from the American Psychological Association

Experimental Condition	P(Reaction) =
Illegitimate Intruder	
Low Social Obligation	0.366
High Social Obligation	0.600
Legitimate Intruder (Acts as Friend)	
Low Social Obligation	0.100
High Social Obligation	0.133

greater the probability of direct verbal and physical reaction by that group member. This is an area of study that could go further in clarifying the severity of an individual's response as a function of increased dissonance. To continue, beginning with the two illegitimate experimental conditions, the probability of the queue member of interest reacting when no one is behind him or her is 0.366, compared to 0.6 when there are two people behind the member of interest. We have observed that Fechner's law provides an excellent representation of how, in the crowd-gathering experiment, the number of confederates looking up influences the reaction probability of an observer. It has also been argued (Dehaene et al., 2003, 2008) that the human brain perceives nonsymbolic numbers on a logarithmic scale, so it may be conjectured that some variation to the generalization of Fechner's Law given in Equation 2.2 will apply to the number of queue members behind the member of interest. To use Fechner's Law under exponential encoding as defined in Equation 2.3, and to account for queue members behind the member of interest, an additional concept from cognitive dissonance theory (Festinger, 1957, pp. 183–185) must now be introduced – that of group attraction.

The data showed, again, that the greater the [group member] attraction to the group, the greater was the degree to which the members tried to influence one another. In other words, the greater the magnitude of dissonance, the stronger the attempt to reduce this dissonance by changing the opinion of the person who disagreed.

Leon Festinger (1957, p. 185, with permission from Stanford University Press)

If group attraction is based on strength of belief in the group's social norms, and the person who disagreed now becomes the intruder who is now part of the group, then any uncertainty regarding group attraction must become part of the social queue equation. Letting group attraction be synonymous with group cohesiveness, the more cohesive the group, the greater the pressure on its members to react to dissonant-causing deviations of group social norms. In ad hoc groups formed for a specific purpose, members not familiar with one another may be less certain regarding intra-group attraction. A group-based noise random variable N_{UA} is used to account for this uncertainty of attraction. Let $p, m \in \mathbb{Z}^0$ where m represents the intensity, or number of people behind the queue member of interest, and p the number of intruders. Building on Equation 2.3, there exists a function $x(m)$ in the format of Equation 1.4 such that

$$0 = G_{ES}\left[\overline{es}_o, x(m)\right] \leq G_{ES}\left[\overline{es}_p, x(0)\right]$$

$$= G_{ES}\left(\overline{es}_p\right) \leq G_{ES}\left[\overline{es}_p, x(m)\right] \leq G_{ES}\left[\overline{es}_p, x(\infty)\right] = 1, or$$

$$G_{ES}(\overline{es}_{p,m}) = 1 - e^{-\overline{es}_p \cdot x(m)} = 1 - e^{-\overline{es}_p \cdot \ln\left(e \cdot \frac{m + N_{UA}}{N_{UA}}\right)}.$$

Equation 3.1

Now, instead of number of confederates looking up as in the crowd-gathering experiment, we are interested in number of confederates intruding and the number of members behind the queue member of interest who the intruder cuts in front of. With that distinction, let us try an approach similar to what was done in Chapter 2 using Table 3.1 data.

For illegitimate intrusions, the geometric mean for uncertainty of attraction $\widehat{N}_{UA}$ may be derived using the low and high social obligation cases in Table 3.1 by inverting Equation 3.1, the CDF $G_{ES}(\overline{es}_{P,M})$, such that

$$\frac{Illegitimate\ High\ social\ obligation}{Illegitimate\ Low\ social\ obligation} = \frac{-ln(1-.6)}{-ln(1-.366)} = \frac{0.916}{0.456} =$$

$$2.009 = \frac{\overline{es} \cdot ln\left(\dfrac{1+\widehat{N}_0}{\widehat{N}_0}\right) \cdot ln\left(e \cdot \dfrac{2+\widehat{N}_{UA}}{\widehat{N}_{UA}}\right)}{\overline{es} \cdot ln\left(\dfrac{1+\widehat{N}_0}{\widehat{N}_0}\right) \cdot ln\left(e \cdot \dfrac{0+\widehat{N}_{UA}}{\widehat{N}_{UA}}\right)} = ln\left(e \cdot \dfrac{2+\widehat{N}_{UA}}{\widehat{N}_{UA}}\right).$$

Solving for $\widehat{N}_{UA}$ results in $\widehat{N}_{UA} = 1.147$. The sensation magnitude sample mean $\overline{es}_p$ for this given social situation and set of social norms may be obtained using the illegitimate low social obligation condition where,

$$illegitimate\ (il)\ case\ sensation\ magnitude\ \overline{es}_{il,1,0} = -ln(1-0.366)$$

$$= 0.456.$$

For the two illegitimate (il) cases, using Equation 3.1, we now have,

$$G_{ES}(\overline{es}_{il,1,0}) = 1 - e^{-0.456 \cdot ln(e)} = 0.366,\ and$$

$$G_{ES}(\overline{es}_{il,1,2}) = 1 - e^{-0.456 \cdot ln\left(e \cdot \frac{2+1.147}{1.147}\right)} = 0.6.$$

which are the original values. Since the illegitimate and legitimate cases are for the same social situation and social norms, we will assume for now until more data comes along that group noise caused by uncertainty of attraction in the legitimate case is the same. Now, using the value $\widehat{N}_{UA} = 1.147$, we can determine an average for legitimate high and low social obligation sensation magnitude using Table 3.1 data to obtain a more accurate sample mean for legitimate (le) intrusion sensation magnitude $\overline{es}_{le,1,0}$,

$$(le)\ low\ social\ obligation\ \overline{es}_{le,1,0} = -ln(1-0.1) = 0.105.$$

Converting (le) high to (le) low social obligation;

$$\overline{es}_{le,1,0} = \frac{-ln(1-0.133)}{ln\left(e \cdot \dfrac{2+1.147}{1.147}\right)} = 0.071.$$

For the legitimate case, we then have the average,

$$(le) \ low \ social \ obligation \ \overline{es}_{le,1,0} = \frac{0.105 + 0.071}{2} = 0.088.$$

Therefore, for the two legitimate cases

$$G_{ES}(\overline{es}_{le,1,0}) = 1 - e^{-0.088 \cdot ln(e)}$$

$$= 0.084, which \ is \ within \ the \ 95\% \ credibility \ interval$$

$$(.036, .258).$$

$$G_{ES}(\overline{es}_{le,1,2}) = 1 - e^{-0.088 \cdot ln\left(e \cdot \frac{2+1.147}{1.147}\right)}$$

$$= 0.162, which \ is \ within \ the \ 95\% \ credibility \ interval$$

$$(.054, .298).$$

Given only a single intruder, there is currently insufficient information to calculate unit stimulus sensation magnitude $\overline{es}$ and social noise $\widehat{N}_0$ for either the illegitimate or legitimate case.

With Equation 3.1, representing amplification of social pressure on the queue member of interest to react as a function of members in queue behind him or her, and with $\overline{es}_{il,1,0}$ and $\widehat{N}_{UA}$ derived using Grand Central railroad station data, we are ready to analyze the more complex Milgram et al. (1986) intrusion data, consisting of 17 locational social situations, the use of one or two intruders, various numbers of queue members behind the member of interest, and having from zero to two non-reacting confederate buffers between the intruder and queue member of interest.

3.2 Milgram's "Response to Intrusion Into Waiting Lines": Reaction Summary and Associated Social Situations

Before analyzing Milgram's intrusion experiment, its structure and the social situations from which the data was taken must be clearly understood. Three graduate student reports graded by Dr. Milgram in 1978 were found at Stanley Milgram Papers (MS 1406), Manuscripts and Archives, Yale University Library, along with supplementary raw summary data. The student reports and supplementary data clarify how the experiment was conducted and what was learned. In a few instances, data was found that is otherwise not available in the published article by Milgram et al. (1986). Sadly, few members of the larger experiment team are still alive, and those that are do not have the 126 coding data observation sheets (see example in Appendix A) which would allow for a more straightforward analysis of the overall set of experiments. These data sheets had information on every trial, including location of each observation, date, time, how each queue member reacted as a function of queue position, and how many queue members were in each observed queue. As a result, to address the latter, queueing theory is applied (or abused for the pure at heart) in Appendix A to approximate the actual mean length of the Condition 4 queue. The Condition 4 queue, in 17 of its 23 trials, clearly extends beyond what is available in the raw summary data, which ends at the fifth queue member position behind the point of intrusion. The law of total probability and the chain rule from Bayesian statistics are heavily employed to approximate group statistics that cannot otherwise be evaluated directly due to the absence of coding data. All of this is to say that, as we wade into the math, the approach requires some assumptions which are implemented in a transparent and methodical manner. It is up to the reader to evaluate their merit in trying to do the best with what data exists. To allay any fears, this chapter will not finish with the statement, "more data is needed." It will end with a complete model prototype allowing quantitative evaluation against additional social experiments with a queue-like structure for either disproving or further validating the model as derived. This hopefully begins to address what Thomas Pettigrew was looking for in his reassessment commentary of social psychological theory (Pettigrew, 1986, pp. 169–195).

The upcoming calculations focus predominantly on data contained in both Milgram et al. (1986, p. 685, Table 3.2) and the Stanley Milgram Papers (MS 1406), Manuscripts and Archives, Yale University Library, student experiment reports of Raymond

TABLE 3.2 Original 1978 Raw Summary Reaction Data by Condition and Position. Note that Milgram et al. (1986) Data Takes Precedence When Any Differences are Present

		Condition 1: (+1) - Q(1|0|*)	Condition 1: (+2) - Q(1|1|*)	Condition 1: (+3) - Q(1|2|*)	Condition 2: (+2) - Q(1|1|*) 1 Buffer	Condition 2: (+3) - Q(1|2|*) 1 Buffer	Condition 3: (+3) - Q(1|2|*) 2 Buffers	Condition 4: (+1) - Q(2|0|*)	Condition 4: (+2) - Q(2|1|*)	Condition 4: (+3) - Q(2|2|*)	Condition 5: (+2) - Q(2|1|*) 1 Buffer	Condition 5: (+3) - Q(2|2|*) 1 Buffer	Condition 6: (+3) - Q(2|2|*) 2 Buffers	Condition 6: (+4) - Q(2|3|*) 2 Buffers
Milgram et al (1986) Data	Breakdown of queue member actions as a function of queue condition and queue member position Q(P|N|M). P is the number of simulataneous intruders, N is the number of queue members or confederate non-reactive buffers between member of interest and intruder(s), and M is the number of queue members behind the member of interest.													
	Probability of Member Reaction to Intruder Based on Experimental Data:	0.364	0.143	0.000	0.167	0.000	0.000	0.870	0.435	0.091	0.200	0.000	0.150	0.118
	# of Reactions Divided by Total Observations per Condition and Queue Member of Interest (+[N+1])	8/22	2/14	0/9	4/24	0/15	0/20	20/23	10/23	2/22	4/20	0/15	3/20	2/17
Stanley Milgram Papers - Archives (MS 1406, Box 100, Folder 131), Yale University Library	# Physical Ejection of Intruder (PE)	1						6	2				1	1
	# Verbal Ejection of Intruder (VE)	4	1		1			9	7	3				1
	# Verbal Disapproval to Intruder (VDI)	3						3	1		1			
	# Nonverbal Dissaproval to Intruder (NDI)		1		3			2		1	4		2	
	Reaction Counted Only When Directed at Intruder													
	# Verbal Dissaproval to Others (VDO)	2					1		2		1	1	2	1
	# Nonverbal Disapproval to Others (NDO)		2	1			1			1	2		1	
	# Undirected Dissaproval (UD)	2	0		4	1	1				3	2	4	1
	# No Reaction Detected (NRD)	7	10	8	15	14	17	2	10	17	9	12	10	12
	#Verbal/Nonverbal Approval (VNA)	3			1			1						

Unpublished data obtained from The Stanley Milgram Papers (MS 1406) Manuscripts and Archives, Yale University Library with permission from the Estate of Alexandra Milgram. Where Stanley Milgram Paper raw data differs slightly, precedence given to final data in Milgram et al (1986).

Toledo, David Nemiroff, and Christina Taylor who were graduate students under Stanley Milgram in 1978. Also used from the Stanley Milgram Papers (MS 1406) are four pages of rough notes written in preparation for the line intrusion experiment by then graduate student Joyce Wackenhut, raw summary queue member line position reaction data for each of the six conditions (individual trial data not available), and summary data regarding queue locations for each of the six conditions. After presenting pertinent summary data from the experiment as a function of experiment condition, the order of analysis taken for each of the six conditions is performed to support derivation of necessary values and additional analytic tools. Analysis begins with Condition 1, composed of 22 trials from a total of eight different queue locations. In Condition 1, a single confederate intruder is observed entering a line at a position (termed queue position zero) having typically five or less queue members after the intrusion point. Each queue member position after the intrusion point is labeled as position $(+1)$, $(+2)$, $(+3)$, $(+4)$, and $(+5)$. Queue members forward of the intrusion point are labeled as positions (-1), (-2), (-3), and (-4) and are not analyzed given the additional variables which cannot be accounted for such as did they even notice. Both Milgram et al. (1986) and Schmitt et al. (1992) provide more detailed discussion on the latter but neither reach a definitive conclusion.

Table 3.2 is consolidated from the original 1978 raw summary data obtained from Stanley Milgram Papers (MS 1406). Any reaction by the queue member of interest directed at the intruder is counted as a reaction. For example, "Condition 1: $(+1) - Q(1 \,|\, 0 \,|\, *)$" indicates number of reactions from the Condition 1 set of observations having a single intruder "1," with no members "0" between the intruder and the member of interest who is therefore at the $(+1)$ position, and with various numbers "*" of queue members behind the member of interest. In this case, of the 22 observations, 8 of the 22 queue members at the $(+1)$ position react directly to the intruder. As another example, Condition 4: $(+3) - Q(2 \,|\, 2 \,|\, *)$ indicates probability of reaction resulting from two simultaneous intruders "2," by queue member position $(+3)$ (i.e., two intermediate queue members fill the $(+1)$ and $(+2)$ positions in front of the position of interest), with various numbers "*" of queue members behind the position of interest. Of the 22 trials where two intruders simultaneously intrude at position (0), the $(+3)$ position queue members with two queue members between themselves and the two intruders are observed to react twice out of 22 observations. Looking at Table 3.2 for Condition 4: $(+1)$, it is readily apparent there are more reactions in this condition than for Condition 1: $(+1)$. As importantly, the severity of reactions in Condition 1: $(+1)$ is mostly verbal disapproval or verbal ejection. In Condition 4: $(+1)$, the severity of member reaction moves up predominantly to verbal and physical ejection of the intruders. This could be due to the number of queue members behind the queue member of interest, it could be due to the number of intruders, or it could be due to the social situation (i.e., importance or limited availability of the queue resource commodity). As we will discover shortly, it is due to all three. Basically, as sensation magnitude increases, so does member aggression. This supports Axiom 1 and the quote by Festinger (1957, p. 18, with permission from Stanford University Press), "*The strength of the pressures to reduce the dissonance is a function of the magnitude of the dissonance.*" Something similar is obviously occurring in the queue as in the crowd data experiment. As seen in the crowd-gathering data, the larger the crowd that deviates from a social norm, the greater the dissonance felt by the observer and the greater the pressure on average to look up and reduce the dissonance.

Indication has been provided that the queue's social situation plays a role in the queue member's sensation magnitude resulting from an intrusion. This supports the person–situation interaction equation (Lewin, 1997, p. 187; Burnes and Cooke, 2013, pp. 412–413). To lay the foundation for this claim, locations and situations of queues are listed in Table 3.3 by Condition. Focus should be placed on the social situation, not the location. For instance, we may ask if waiting in line for a train ticket, having a relatively strict time schedule, is the same as waiting in line for a snow cone on our way

TABLE 3.3 Queue Experiment Locations and Frequency of Use by Condition

Locations that Queue Intrusion Data was Observed From:	Milgram et al (1986) Condition 1	Milgram et al (1986) Condition 2	Milgram et al (1986) Condition 3	Milgram et al (1986) Condition 4	Milgram et al (1986) Condition 5	Milgram et al (1986) Condition 6	Schmitt et al (1992) – Illegitimate	Schmitt et al (1992) – Legitimate
Bank (86)	2	0	2	0	1	0	0	0
Choc Full of Nuts (42)	0	0	0	0	1	0	0	0
Gyros Restaurant (42)	0	0	1	0	0	0	0	0
Health Food Store (GC)	0	0	1	0	0	0	0	0
Information (GC)	1	1	0	2	1	0	0	0
Investment (GC)	2	0	2	1	0	0	0	0
Passport (50)	1	0	1	0	0	0	0	0
Post Office (43)	0	0	0	0	4	0	0	0
Snow Cone Store (GC)	0	1	0	0	0	0	0	0
Zaro's Restaurant (34)	0	0	1	0	0	0	0	0
Token Booth (GC)	0	3	3	0	0	0	0	0
Ticketron (GC)	4	0	2	1	1	0	0	0
Low Unit Stimulus Sensation Magnitude Categories (es_L) - White								
Bus Tickets (PA)	0	0	0	0	0	7	0	0
Bus Boarding (PA)	0	0	0	0	0	8	0	0
Airport (L)	1	0	2	0	0	0	0	0
Ticket Rail (GC)	8	16	3	18	6	4	30	30
Off-track Betting (OTB) (GC)	3	3	2	1	6	1	0	0
High Unit Stimulus Sensation Magnitude Categories (es_H) - Gray Shaded								
Total Number of Observations:	22	24	20	23	20	20	30	30
A - # Hypothesized High Unit Stimulus Sensation Magnitude	12	19	7	19	12	20	30	0
B - # Hypothesized Low Unit Stimulus Sensation Magnitude	10	5	13	4	8	0	0	30
For Weighting:								
Pr(A) =	0.545	0.792	0.350	0.826	0.600	1.000	1.000	0.000
Pr(B) =	0.455	0.208	0.650	0.174	0.400	0.000	0.000	1.000

(L) - LaGuardia, TWA Boarding Line   (PA) - Port Authority   (GC) - Grand Central Terminal

(#) - Street Locations in Manhatten City, NY   (OTB) - Off-track betting parlor

Note * : Graduate student David Nemiroff indicates for Condition 4, that two of the three non-reactions were at the GC Information Booth and that the Information Booth queue was amorphous (i.e. not well structured as a queue).

Data obtained from the Stanley Milgram Papers (MS 1406) Manuscripts and Archives, Yale University Library with permission from the Estate of Alexandra Milgram.

home. It is hypothesized here that limited resources (i.e., time to get the desired train before it departs, selection of a "good" seat when boarding a bus, time available to make a bet on your favorite horse before the race starts) are of greater importance and hence result in greater extent social deviation caused by an intruder than compared to queues with abundant or relatively unlimited resources (i.e., time, snow cones, stamps at the post office, passport ID, etc.). Those queues proposed as higher unit stimulus sensation magnitude queues include the five gray shaded locations of Table 3.3 which include the bus terminal, airport, ticket rail, and off-track betting locations. On the other hand, the 12 queue locations above the gray shaded area in Table 3.3 support queues waiting for relatively unlimited resources in a more relaxed setting. For example, Condition 1 encompasses eight different locations, where 12 observations are in potentially high and 10 are in potentially low-unit stimulus sensation magnitude situations. Alternatively, 18 of 23 queue observations for Condition 4 are in the potentially high-unit stimulus sensation magnitude ticket rail queue at Grand Central Terminal, the same location as the data from Schmitt et al. (1992). With this background, we are now ready to begin quantitative analysis of the data.

3.3 Milgram's "Response to Intrusion Into Waiting Lines": Analysis and Model Development

We begin here with the tools developed and values derived from Schmitt et al. (1992), which allow us to begin quantitative evaluation of the Milgram et al. (1986) experiment. Using results derived from Schmitt et al. (1992) for $\overline{es}_{il,1}$ and $\hat{N}_{UA}$, the single intruder sensation magnitude value for low sensation magnitude queues is derived using Condition 1: $(+1) - Q(1\,|\,0\,|*)$. This in turn should support derivation of the two-intruder high- and low-unit sensation magnitude sample mean using Condition 4: $(+1) - Q(2\,|\,0\,|*)$. If Corollary 2.3 is met, the results from these two conditions and associated situations may then be applied to the remaining queue conditions to improve sample mean values and validate model predictions regarding impact of queue members between the intruder(s) and position of interest. Once complete, a spreadsheet incorporating the final model and all conditions simultaneously is used to fine tune results using the mean least squares method. With that complete, comparison between theoretical predictions versus empirical results is made against the 95 percent credibility intervals based on empirical results.

3.3.1 Condition 1: (+1)-Q(1|0|M) Probability of Reaction Calculation

Table 3.4 represents data from Condition 1 of Milgram et al. (1986). The associated analysis focuses on this data. For now, define N as the number of queue members between the intruder(s) and the position of interest. The number of queue members $m \in M$ behind the position of interest is found in the macro variable $Q(P\,|\,N\,|\,M)$ of Table 3.4. This variable also identifies the associated fraction of trials in which this configuration exists as a function of queue member position. To understand how the data was obtained, consider $Q(1\,|\,0\,|\,0)$ at position $(+1)$. There were 22 trials observed, yet in position $(+2)$ there were only 14 trials having members in that position. That

TABLE 3.4 Condition 1 Empirical Data

$Q(P\|N\|M) = Q(P$ intruders | N queue members between intruder and position of interest | M queue members behind queue position of interest) = Position Probability of Occurrence				
Position (-1) ⬇	⬅ Position (-2)	⬅ Position (-3)	⬅ Position (-4)	⬅ Front of Queue
5/22	1/22	0/21	0/5	# React /# Opportunities
🧍 **1 Intruder** **(Condition 1)** ⬇	<u>Notes</u>: 1) Appendix A has k=1 and l_{100}=64.9: theoretical L = 1.8 ≈ documented M_{mean} = 1.7 in the Q(1|0|*) configuration: No extrapolation for "M" necessary.			
Position (+1) ⬇	⟹ Position (+2)	⟹ Position (+3)	⟹ Position (+4)	⟹ Position (+5)
8/22	2/14	0/9	0/7	0/7
$Q(1\|0\|4) = 7/22$	$Q(1\|1\|3) = 7/14$	$Q(1\|2\|2) = 7/9$	$Q(1\|3\|1) = 7/7$	$Q(1\|4\|0) = 7/7$
$Q(1\|0\|3) = 0/22$	$Q(1\|1\|2) = 0/14$	$Q(1\|2\|1) = 0/9$	$Q(1\|3\|0) = 0/7$	Milgram et al (1986)
$Q(1\|0\|2) = 2/22$	$Q(1\|1\|1) = 2/14$	$Q(1\|2\|0) = 2/9$	supplemented by unpublished data obtained from	
$Q(1\|0\|1) = 5/22$	$Q(1\|1\|0) = 5/14$	The Stanley Milgram Papers (MS 1406), Series No. IV, Box No. 100, Folder No. 131		
$Q(1\|0\|0) = 8/22$	Manuscripts and Archives, Yale University Library with copyright permission from the Estate of Alexandra Milgram.			

means that there were $22 - 14 = 8$ trials out of 22 with zero queue members behind the $(+1)$ position of interest. For position $(+3)$, there were 9 trials out of 22 where the $(+3)$ position had a member in it. Since the $(+2)$ position had 14 trial observations with a member in it, that means there were $14 - 9 = 5$ trials where there were no queue members behind the $(+2)$ position (i.e., $Q(1\|1\|0) = 5/14$). The same process is used to obtain the remaining values.

Let us start with the proposed sample mean sensation magnitude value $\overline{es}_{il,1} = 0.456$ for illegitimate intrusions and uncertainty of attraction $\widehat{N}_{UA} = 1.147$ as was derived from the Schmitt et al. (1992) data. Since the illegitimate intrusion data was obtained at the Grand Central Terminal ticket counter, the result applies to high-unit stimulus sensation magnitude queues. Define $\overline{es}_{H,1,0} = \overline{es}_{H,1} = 0.456$ as the sensation magnitude value for high (H) unit sensation magnitude queues with one intruder, zero members between the intruder and position of interest, and zero members behind the position of interest. Similarly, define the low (L) queue unit sensation magnitude value as $\overline{es}_L$. Given $\overline{es}_{H,1} = 0.456$, $\widehat{N}_{UA} = 1.147$, with Condition 1: $(+1)$ results, we are now able to derive $\overline{es}_{L,1,m}$ using:

$$\overline{es}_{L,1,m} = \overline{es}_{L,1} \cdot ln\left(e \cdot \frac{m + \widehat{N}_{UA}}{\widehat{N}_{UA}} \right)$$

Together, using the chain rule and the law of total probability for Condition 1: $(+1)$, we obtain the probability of reaction at the $(+1)$ queue position $P(R_{+1})$ for unit sensation magnitude $\overline{es}_I, I \in \{L, H\}$. The probability of $P(\overline{es}_I)$ is obtained from Table 3.3. Using the probability of queue member position $P(m) = Q(1\|0\|m)$ from Table 3.4 where $m \in M = \{0, 1, 2, 3, 4\}$,

$$P(R_{+1}) = \sum_{I \in \{L,H\}} \sum_{m \in M} P\left(ES \leq \overline{es}_{I,1,m}, \overline{es}_I, m \right)$$

$$= \sum_{I \in \{L,H\}} \sum_{m \in M} G_{ES}\left(\overline{es}_{I,1,m} \right) \cdot P\left(\overline{es}_I \| m \right) \cdot P\left(m \right)$$

$$= \sum_{I \in \{L,H\}} \sum_{m \in M} \left[1 - e^{-\overline{es}_I ln\left(\frac{1+\tilde{N}_0}{\tilde{N}_0}\right) \cdot ln\left(e \cdot \frac{m+\tilde{N}_{UA}}{\tilde{N}_{UA}}\right)} \right]$$

$$\cdot P\left(\overline{es}_I \mid m\right) \cdot P(m).$$

For our purposes $P(R_{+1})$, the probability of reaction at the (+1) position, may be stated as

$$P(R_{+1}) = \sum_{I \in \{H,L\}} \sum_{m \in M} P\left[ES \le \overline{es}_{I,1,m}, \overline{es}_I, Q(1|0|m) \right]$$

$$= \sum_{I \in \{H,L\}} \sum_{m \in M} G_{ES}\left(\overline{es}_{I,1,m}\right) \cdot P\left[\overline{es}_I \mid Q(1|0|m) \right]$$

$$\cdot P\left[Q(1|0|m) \right]$$

The Condition 1:(+1) probability of reaction may now be calculated using Tables 3.3 and 3.4. Since no remaining experimental data exists to do otherwise, we will necessarily assume throughout this and subsequent analyses that the proportion of high and low unit stimulus sensation magnitude queues is independent of m, implying $P\left[\overline{es}_I \mid Q(1|0|m) \right] = P\left[\overline{es}_I \right]$.

$$P\left[ES \le \overline{es}_{I,1,m}, \overline{es}_H, Q(1|0|0) \right] = \left(1 - e^{-0.456}\right) \cdot \left(\frac{12 \, \overline{es}_H \, Trials}{22 \, Trials}\right) \cdot \left(\frac{8}{22}\right)$$

$$= 0.073$$

$$P\left[ES \le \overline{es}_{I,1,m}, \overline{es}_H, Q(1|0|1) \right] = \left(1 - e^{-0.456 \cdot ln\left(e \cdot \frac{1+1.147}{1.147}\right)}\right) \cdot \left(\frac{12}{22}\right) \cdot \left(\frac{5}{22}\right)$$

$$= 0.065$$

$$P\left[ES \le \overline{es}_{I,1,m}, \overline{es}_H, Q(1|0|2) \right] = \left(1 - e^{-0.456 \cdot ln\left(e \cdot \frac{2+1.147}{1.147}\right)}\right) \cdot \left(\frac{12}{22}\right) \cdot \left(\frac{2}{22}\right)$$

$$= 0.030$$

$$P\left[ES \leq \overline{es}_{I,1,m}, \overline{es}_H, Q(1|0|3)\right] = \left(1 - e^{-0.456 \cdot ln\left(e \cdot \frac{3+1.147}{1.147}\right)}\right) \cdot \left(\frac{12}{22}\right) \cdot \left(\frac{0}{22}\right)$$

$$= 0.0$$

$$P\left[ES \leq \overline{es}_{I,1,m}, \overline{es}_H, Q(1|0|4)\right] = \left(1 - e^{-0.456 \cdot ln\left(e \cdot \frac{4+1.147}{1.147}\right)}\right) \cdot \left(\frac{12}{22}\right) \cdot \left(\frac{7}{22}\right)$$

$$= 0.118$$

To match $P(R_{+1}) = \frac{8}{22} = 0.364$ from Condition 1: (+1), a spreadsheet calculation was done to find that $\overline{es}_{L,1,0} = 0.114$. Using $\overline{es}_{L,1,0} = 0.114$, we continue with

$$P\left[ES \leq \overline{es}_{I,1,m}, \overline{es}_L, Q(1|0|0)\right] = \left(1 - e^{-0.114}\right) \cdot \left(\frac{10\,\overline{es}_L\,Trials}{22\,Trials}\right) \cdot \left(\frac{8}{22}\right)$$

$$= 0.018$$

$$P\left[ES \leq \overline{es}_{I,1,m}, \overline{es}_L, Q(1|0|1)\right] = \left(1 - e^{-0.114 \cdot ln\left(e \cdot \frac{1+1.147}{1.147}\right)}\right) \cdot \left(\frac{10}{22}\right) \cdot \left(\frac{5}{22}\right)$$

$$= 0.017$$

$$P\left[ES \leq \overline{es}_{I,1,m}, \overline{es}_L, Q(1|0|2)\right] = \left(1 - e^{-0.114 \cdot ln\left(e \cdot \frac{2+1.147}{1.147}\right)}\right) \cdot \left(\frac{10}{22}\right) \cdot \left(\frac{2}{22}\right)$$

$$= 0.008$$

$$P\left[ES \leq \overline{es}_{I,1,m}, \overline{es}_L, Q(1|0|3)\right] = \left(1 - e^{-0.114 \cdot ln\left(e \cdot \frac{3+1.147}{1.147}\right)}\right) \cdot \left(\frac{10}{22}\right) \cdot \left(\frac{0}{22}\right)$$

$$= 0.0$$

$$P\left[ES \leq \overline{es}_{I,1,m}, \overline{es}_L, Q(1|0|4)\right] = \left(1 - e^{-0.114 \cdot ln\left(e \cdot \frac{4+1.147}{1.147}\right)}\right) \cdot \left(\frac{10}{22}\right) \cdot \left(\frac{7}{22}\right)$$

$$= 0.036$$

From this, and accounting for round-off error,

$$P(R_{+1}) = \sum_{I \in \{H,L\}} \sum_{m \in M} P\left[ES \le \overline{es}_{I,1,m}, \overline{es}_I, Q(1\,|\,0\,|\,m) \right]$$

$$= \sum_{I \in \{H,L\}} \sum_{m \in M} G_{ES}\left(\overline{es}_{I,1,m} \right) \cdot P\left(\overline{es}_I \right) \cdot P\left[Q(1\,|\,0\,|\,m) \right] = 0.364$$

Condition 1: (+1) has now been used to systematically determine a sample mean for the low sensation magnitude value $\overline{es}_{L,1,0} = 0.114$. That there are only two categories, low- and high-unit stimulus sensation magnitude queues, is probably overly simplistic, but we need to begin somewhere, and having a dividing line between limited resource (ticket and time) and perceptually unlimited resource is a good starting point.

3.3.2 Condition 4 – Probability of Reaction Calculation for Q(2|0|M)

To find the sensation magnitude value for a New York City queue with two intruders, we begin with Condition 4: (+1). From Schmitt et al. (1992), the evaluation of the sample-mean sensation magnitude for Grand Central Terminal using 60 trials was $\overline{es}_{H,1,0} = 0.456$ with $\widehat{N}_{UA} = 1.147$. From Section 3.3.1, using Condition 1: (+1) it was found that

$$\overline{es}_{L,1,0} = \overline{es}_L \cdot ln\left(\frac{1+\widehat{N}_0}{\widehat{N}_0} \right) = \frac{\overline{es}_H \cdot ln\left(\dfrac{1+\widehat{N}_0}{\widehat{N}_0} \right)}{4.0} \; so \; \overline{es}_L = \frac{\overline{es}_H}{4.0},$$

and in similar manner $\overline{es}_{L,2,0} = \dfrac{\overline{es}_{H,2,0}}{4.0}.$

With Condition 4: (+1) results provided in Table 3.5, there is enough information now to solve for $\overline{es}_{H,2,0}$ in a New York City high-unit stimulus sensation magnitude queue situation using data from Milgram et al. (1986). The approach taken, using the same process as for Condition 1: (+1), is simple. Since the low-unit stimulus sensation magnitude queue is one-quarter the value of the high unit stimulus sensation magnitude queue, all that needs to be done is to find the value for $\overline{es}_{H,2,0}$, using a spreadsheet, such that after simultaneous summation of the equations, the result $P(R_{+1}) = 0.870$ is reached. Once the value for $\overline{es}_{H,2,0}$ is derived, if Corollary 2.3 is met, then the geometric mean $\widehat{N}_0$ and sample mean value $\overline{es}\left(\widehat{N}_0 \right)$ may be derived, given values for both $\overline{es}_{H,2,0}$ and $\overline{es}_{H,1,0}$ have been successfully determined.

Before proceeding, the "Notes" section of Table 3.5 should be reviewed along with Appendix A to understand the rationale and mathematics behind approximating the actual Condition 4: (+1) queue length in which 17 trials had $m \in M, m \ge 4$ per the summary data. As with Condition 1, let $m \in M = \{0,1,2,3,11.3\}$ represent the elements of the set M containing the various number of queue members behind the queue member of interest.

TABLE 3.5 Condition 4 Empirical Data

$Q(P\|N\|M) = Q(P$ intruders | N queue members between intruder and position of interest | M queue members behind queue position of interest$) = $ Position Probability of Occurrence				
Position (-1) ⬇	⬅ Position (-2)	⬅ Position (-3)	⬅ Position (-4)	⬅ Front of Queue
3/23	1/23	0/22	0/10	# React /# Opportunities
2 Intruders (Condition 4)	**Notes**: 1) Appendix A has k=10 and l_{100}=95.2: theoretical M = 11.3 >> documented M_{mean} = 3.4 in the $Q(2\|0\|*)$ configuration: Changed $Q(2\|0\|4)$ to $Q(2\|0\|11.3)$. 2) Original 1978 raw summary data and two graduate student reports indicate 10/22 at (+2), 4/22 at (+3), and 0/17 at the (+4) positions - differing with Milgram et al (1986) (i.e. 10/23, 2/22, 0/20). Milgram et al (1986) takes precedence.			
Position (+1) ⬇	➡ Position (+2)	➡ Position (+3)	➡ Position (+4)	➡ Position (+5)
20/23	10/23	2/22	0/20	0/17
$Q(2\|0\|11.3)$ = 17/23	$Q(2\|1\|10.3)$ = 17/23	$Q(2\|2\|9.3) = 17/22$	$Q(2\|3\|8.3) = 17/20$	$Q(2\|4\|7.3) = 17/17$
$Q(2\|0\|3) = 3/23$	$Q(2\|1\|2) = 3/23$	$Q(2\|2\|1) = 3/22$	$Q(2\|3\|0) = 3/20$	Milgram et al (1986)
$Q(2\|0\|2) = 2/23$	$Q(2\|1\|1) = 2/23$	$Q(2\|2\|0) = 2/22$	supplemented by unpublished data obtained from	
$Q(2\|0\|1) = 1/23$	$Q(2\|1\|0) = 1/23$	The Stanley Milgram Papers (MS 1406), Series No. IV, Box No. 100, Folder No. 131		
$Q(2\|0\|0) = 0/23$	Manuscripts and Archives, Yale University Library with copyright permission from the Estate of Alexandra Milgram.			

By spreadsheet, based on the following equations, it is found that $\overline{es}_{H,2,0} = \overline{es}_{H,2}$ $= 1.005$ and $\overline{es}_{L,2,0} = \overline{es}_{L,2} = \dfrac{\overline{es}_{H,2}}{4.0} = 0.251$. As a function of $\overline{es}_{H,2}$, the equations used in the spreadsheet supporting a simultaneous solution are

$$P\left[ES \le \overline{es}_{I,2,m}, \overline{es}_H, Q(2|0|0) \right]$$

$$= \left(1 - e^{-1.005}\right) \cdot \left(\frac{19\,\overline{es}_H\ Trials}{23\ Trials} \right) \cdot \left(\frac{0}{23} \right) = 0.0$$

$$P\left[ES \le \overline{es}_{I,2,m}, \overline{es}_H, Q(2|0|1) \right] = \left(1 - e^{-1.005 \cdot ln\left(e \cdot \frac{1+1.147}{1.147} \right)} \right) \cdot \left(\frac{19}{23} \right) \cdot \left(\frac{1}{23} \right)$$

$$= 0.029$$

$$P\left[ES \le \overline{es}_{I,2,m}, \overline{es}_H, Q(2|0|2) \right]$$

$$= \left(1 - e^{-1.005 \cdot ln\left(e \cdot \frac{2+1.147}{1.147} \right)} \right) \cdot \left(\frac{19}{23} \right) \cdot \left(\frac{2}{23} \right) = 0.062$$

$$P\left[ES \le \overline{es}_{I,2,m}, \overline{es}_H, Q(2|0|3) \right]$$

$$= \left(1 - e^{-1.005 \cdot ln\left(e \cdot \frac{3+1.147}{1.147} \right)} \right) \cdot \left(\frac{19}{23} \right) \cdot \left(\frac{3}{23} \right) = 0.097$$

$$P\left[ES \le \overline{es}_{I,2,m}, \overline{es}_H, Q(2|0|11.3) \right]$$

$$= \left(1 - e^{-1.005 \cdot ln\left(e \cdot \frac{11.3+1.147}{1.147} \right)} \right) \cdot \left(\frac{19}{23} \right) \cdot \left(\frac{17}{23} \right) = 0.590$$

$$P\left[ES \le \overline{es}_{I,2,m}, \overline{es}_L, Q(2|0|0) \right]$$

$$= \left(1 - e^{-0.251} \right) \cdot \left(\frac{4 \, \overline{es}_L \; Trials}{23 \, Trials} \right) \cdot \left(\frac{0}{23} \right) = 0.0$$

$$P\left[ES \le \overline{es}_{I,2,m}, \overline{es}_L, Q(2|0|1) \right]$$

$$= \left(1 - e^{-0.251 \cdot ln\left(e \cdot \frac{1+1.147}{1.147} \right)} \right) \cdot \left(\frac{4}{23} \right) \cdot \left(\frac{1}{23} \right) = 0.003$$

$$P\left[ES \le \overline{es}_{I,2,m}, \overline{es}_L, Q(2|0|2) \right]$$

$$= \left(1 - e^{-0.251 \cdot ln\left(e \cdot \frac{2+1.147}{1.147} \right)} \right) \cdot \left(\frac{4}{23} \right) \cdot \left(\frac{2}{23} \right) = 0.006$$

$$P\left[ES \le \overline{es}_{I,2,m}, \overline{es}_L, Q(2|0|3) \right]$$

$$= \left(1 - e^{-0.251 \cdot ln\left(e \cdot \frac{3+1.147}{1.147} \right)} \right) \cdot \left(\frac{4}{23} \right) \cdot \left(\frac{3}{23} \right) = 0.010$$

$$P\left[ES \le \overline{es}_{I,2,m}, \overline{es}_L, Q(2|0|11.3) \right]$$

$$= \left(1 - e^{-0.251 \cdot ln\left(e \cdot \frac{11.3+1.147}{1.147} \right)} \right) \cdot \left(\frac{4}{23} \right) \cdot \left(\frac{17}{23} \right) = 0.074$$

Assuming homogeneity of high- and low-dissonance magnitude queues as before and accounting for round-off error in the aforementioned,

$$P(R_{+1}) = \sum_{I \in \{H,L\}} \sum_{m \in M} P\left[ES \leq \overline{es}_{I,2,m}, \overline{es}_I, Q(2|0|m) \right]$$

$$= \sum_{I \in \{H,L\}} \sum_{m \in M} G_{ES}\left(\overline{es}_{I,2,m} \right) \cdot P\left(\overline{es}_I \right) \cdot P\left[Q(2|0|m) \right] = 0.870$$

Unfortunately, $\frac{\overline{es}_{H,2}}{\overline{es}_{H,1}} = \frac{1.005}{0.456} = 2.2 > \frac{2}{1}$. Hence, as per Corollary 2.3, there is no feasible solution for $\widehat{N}_0$ and hence for $\overline{es}\left(\widehat{N}_0 \right)$ using Condition 4: (+1) data. It is worth mentioning that had one less (+1) reaction occurred from those 23 observations, the resultant sensation magnitude would have been $\overline{es}_{H,2} = 0.81$, and Corollary 2.3 would have been met.

Experiments take time to collect data, and for this experiment, the stress placed on the graduate students as intruders was significant – in some cases leading to feelings of nausea right before the intrusion. More data for this Condition would have reduced the credibility interval and should have brought the sensation magnitude value back within the Corollary 2.3 required region allowing for a feasible solution of $\widehat{N}_0$. But we should be grateful for what we have, and Condition 4: (+2) data will be extremely useful later.

Returning to our current problem, to find a feasible solution for $\widehat{N}_0$ and hence for $\overline{es}\left(\widehat{N}_0 \right)$, the only remaining option is to derive $\overline{es}_{H,2}$ from Conditions 5 and/or 6 and determine if either or both meet Corollary 2.3 requirements. To do so, we must first derive the impact of the nonreactive buffer for Condition 2 using $\overline{es}_{H,1} = 0.456$, which is supported by 60 observations from Schmitt et al. (1992) and 12 observations from the Condition 1: (+1) data.

3.3.3 Impact of Nonreacting (NR) Confederate Buffers: Deriving z(|UI|) Using Conditions 2: (+2), 5: (+3), and 6: (+3)

Condition 2: (+2), 5: (+2), and 6: (+3) queues are chosen for the following reasons:

1. They are consistent with the raw data from Stanley Milgram Papers (MS 1406) and Milgram et al. (1986).
2. No extrapolation is required to determine the number of members behind the position of interest as was needed in Condition 4.
3. All queue members between the intruder(s) and position of interest are nonreactive confederate student buffers.
4. Each of the three conditions has an empirical probability of reaction greater than zero.

So far, intermediate queue members between the intruder(s) and the position of interest have not been considered. For $p \in P$ and $m \in M$, let us now replace N by defining

UI as the set containing ordered queue position reaction R_i or non-reaction NR_i information for each intermediate queue member at queue position $(+i) = 1, 2, \ldots, |UI|$, where $N = |UI|$ is the cardinality of UI. Equation 3.1 currently considers the null set $UI = \{\ \} \equiv \varnothing \Rightarrow |UI| = 0$ in which case $P(R_{+|UI|+1})$ indicates the $(+1)$ position of interest probability of reaction or $P(R_{+1})$.

Now we will develop Equation 3.1 further to address $|UI| > 0$. By observation of Table 3.2, probability of reaction decreases with each succeeding member position of interest after the intrusion point, regardless of whether each intermediate queue member reacts or does not react. Focusing only on non-reacting NR_i intermediate queue members for the moment, for some function $z(|UI|)$, $p > 0$, and $m \geq 0$, we observe $G_{ES}\left[\overline{es}_{p,m}, z(|UI|)\right]$ decreases as $|UI|$ increases. Furthermore, if $|UI| = 0$, then $G_{ES}\left[\overline{es}_{p,m}, z(0)\right] = G_{ES}\left(\overline{es}_{p,m}\right)$. It will be assumed for an infinite number of intermediate queue members $|UI| \to \infty$ that probability of reaction in a queue at the position of interest tends to zero such that $G_{ES}\left[\overline{es}_{p,m}, z(|UI|)\right] = 0 \ as \ |UI| \to \infty$. Finally, if there are no intruders (i.e., $p = 0$), probability of reaction $G_{ES}\left[\overline{es}_{0,m}, z(|UI|)\right] = 0$. Then for any given nonnegative integers $p, |UI|,$ *and* m, constraints on $G_{ES}\left[\overline{es}_{p,m},\right.$ $\left. z(|UI|)\right]$ are summarized as

$$
\begin{aligned}
0 &= G_{ES}\left[\overline{es}_{0,m}, z(|UI|)\right] \\
&= \lim_{i \to \infty} G_{ES}\left[\overline{es}_{p,m}, z\left(|\{NR_1, NR_2, \ldots, NR_i\}|\right)\right] \\
&\leq G_{ES}\left[\overline{es}_{p,m}, z(|UI|)\right] \leq G_{ES}\left[\overline{es}_{p,m}, z(0)\right] \\
&= G_{ES}\left(\overline{es}_{p,m}\right).
\end{aligned}
$$

In words, the subject of interest is obtaining information from each intermediate position group member between him or her and the intruder. Those intermediate members who react indicate to the subject of interest that deviation from a social norm likely occurred. Their reaction, as judged by the next further observing intermediate member in queue, will range from either being sufficient or insufficient, the latter possibly warranting some further level of reaction. Those intermediate members who do not react indicate to the subject of interest that the social deviation was acceptable to them, or at least not sufficient to warrant a reaction as far as they were concerned. Each reaction or non-reaction by an intermediate queue member may be viewed as an increment of dissonance-reducing information made use of by the subject of interest to reduce uncertainty prior to making the choice of whether or not to react. When uncertainty of choice surrounding an event exists, it is hypothesized that dissonance-reducing information directed at relieving the uncertainty will reduce the event-induced dissonance.

When social events occur that are not congruent with how we believe people should act, based on our social norms for the given situation, then we experience emotional discomfort and associated dissonance. If the social event is ambiguous within the given social situation, then we experience uncertainty and must make a choice regarding what action or nonaction provides the best opportunity to reduce the associated dissonance. Information is sought to make the best choice, and as Festinger (1957, pp. 177–202) indicates, the social group offers the best means for dissonance reduction by providing the necessary support from other group members who share similar social norms. Since the information we seek as per Axiom 1 is to reduce dissonance for the given socially deviant event and social situation, those in our particular social group, sharing common social norms, would offer the most reliable and pertinent sources of information concerning how to interpret the event and whether sufficient action has been taken.

We are focused on the ambiguous socially deviant event which creates uncertainty in how to respond to the given situation (Clark and Word, 1972). Reaction is one way to gain information and reduce dissonance; and observation of others in the group is a lower risk approach (see Axiom 2, Chapter 1). As a result, when not in the (+1) position, dissonance-reducing information pertinent to the generated uncertainty is required and probably desired when choosing which action increases the likelihood of successful dissonance reduction.

Using the concepts and nomenclature from Klir (2006, p. 7), the amount of A Priori uncertainty associated with the deviant social event occurring within a group having established and accepted social norms is U_{Pr}. After uncertainty-based dissonance-reducing information associated with reducing U_{Pr} has been obtained by the subject of interest through observing the non-reactions NR_i of intermediate group members, some residual A Posteriori uncertainty will likely remain. Denote this remaining A Posteriori uncertainty as U_{Po}. Define the uncertainty-based dissonance-reducing information set as $UI = \left\{ \emptyset_0, NR_1, NR_2, \ldots, NR_{|UI|} \right\}$ in which $UI_i \in UI$ such that $UI_0 = \emptyset$, $UI_1 = NR_1$, and so on. Then $\bigcup_{i=0}^{|UI|} UI_i$ as the union of uncertainty-based information obtained through observing $|UI|$ intermediate member non-reactions, keeping in mind that $|\emptyset| = 0$. The total amount of uncertainty-based information gained as a function of uncertainty reduction is given by,

$$\left(A\ Priori\ uncertainty \right) - \left(A\ Posteriori\ uncertainty \right) = U_{Pr} - U_{Po}$$

$$= \bigcup_{i=0}^{|UI|} UI_i.$$

This is equivalent to (Smith et al., 2015, pp. 99(d) & 106(e)),

$$U_{Po} = U_{Pr} - \bigcup_{i=0}^{|UI|} UI_i = U_{Pr} \cap \left(\bigcup_{i=0}^{|UI|} UI_i \right)^C = U_{Pr} \cap \bigcap_{i=0}^{|UI|} UI_i^C.$$

It has recently been demonstrated that attitude change during the decision process begins before a choice is made, not after as traditionally believed, indicating that dissonance reduction and spreading of alternatives are already occurring during the decision process. Furthermore, it appears that the person making the choice seeks positive features that are distinctive to one of the choice alternatives (Kitayama et al., 2013; Jarcho et al., 2011). Festinger also implies that dissonance reduction begins to occur while making the choice (Festinger, 1962, pp.93 & 95; Axiom 1). These studies lead to the argument that uncertainty-based information is required to reduce dissonance caused by an ambiguous socially deviant event. Furthermore, for the queue situation, the most appealing choice appears to be non-reaction to an intrusion since it offers minimal risk of negative consequences. If that is the case, then the subject of interest should be focusing his or her uncertainty-based information search on non-reaction from the intermediate queue members, indicating legitimacy and a reduced need for confrontation with the intruder by the subject of interest. We will shortly observe that both arguments are supported by experimental data.

When the intrusion occurs, there is the initially induced dissonance and resulting need to reduce it. To reduce dissonance in ambiguous situations, and not increase it due to an error (e.g., embarrassment or offense), reliable dissonance-reducing information is needed on which to base a choice regarding whether the dissonance-causing event is legitimate or illegitimate. Before making the choice as to whether to confront the intruder, uncertainty-based information is desired to reduce the situational uncertainty. If the choice ultimately made is in error, then embarrassment (Miller, 2001) and/or justifiable confrontation may ensue. Regarding the (+1) position of interest, uncertainty-based information is limited to any immediately preceding social interaction and/or questioning the intruder as to why he or she entered the queue at that position. The initial uncertainty U_{Pr} prompts the need for information on which to base a choice of reaction or no reaction and thus reduce dissonance. It is therefore argued that U_{Pr} is independent of the uncertainty-based information sought to reduce uncertainty by the subject of interest. It may then be stated for position (+1) that, with no intermediate queue members providing information,

$$P\left(R_{+1}\right) \equiv P\left(U_{Po}\right) = P\left(U_{Pr} \cap \bigcap_{i=0}^{|UI|=0} UI_i^C\right) = P\left(U_{Pr}\right)$$

$$\equiv G_{ES}\left(\overline{es}_{p,m}\right).$$

If infinite (i.e., exhaustive) uncertainty-based information is obtained before making the choice of reaction or no reaction in response to the socially deviant event, then uncertainty has been eliminated so that $U_{Po} \to 0$ as $|UI| \to \infty$. This unlikely real-world situation leads to

$$P\left(U_{Po}\right) = P\left(U_{Pr} \cap \bigcap_{i=0}^{|UI|=\infty} UI_i^C\right) = 0 \ implying \ U_{Pr} = \bigcup_{i=0}^{|UI|=\infty} UI_i.$$

Taking this approach indicates that uncertainty-based information pertains only to the uncertainty created by the social norm deviation. In other words

$$\bigcup_{i=0}^{|UI|} UI_i \subseteq U_{Pr} \ for \ |UI| \ge 0.$$

For the case where $0 < |UI| < \infty$, independent of the member of interest, uncertainty-based information is obtained from intermediate queue members immediately after the intrusion, and in the absence of attitude inoculation,

$$P\left(U_{Po}\right) = P\left(U_{Pr} \cap \bigcap_{i=0}^{|UI|} UI_i^{c}\right) = P\left(U_{Pr}\right) \cdot P\left(\bigcap_{i=0}^{|UI|} UI_i^{c}\right)$$

with the implication that $\lim\limits_{|UI| \to \infty} P\left(\bigcap_{i=0}^{|UI|} UI_i^{C}\right) = 0.$

We already know that if $|UI| = 0 \Rightarrow UI = \{\varnothing_0\}$, where no non-reacting intermediate queue members exist between the intruder(s) and subject of interest, then in the absence of any group uncertainty-based information

$$P\left(\bigcap_{i=0}^{|UI|=0} UI_i^{C}\right) = 1.$$

To put this all together, we start with the Chain Rule,

$$P\left(\bigcap_{i=0}^{|UI|} UI_i^{C}\right) = \prod_{i=0}^{|UI|-1} P\left(UI_{|UI|-i}^{C} \mid \bigcap_{j=0}^{|UI|-(i+1)} UI_j^{C}\right).$$

Viewing each of the $|UI|$ non-reacting intermediate queue members as intensity increments of uncertainty-based information which the subject of interest uses in determining whether to react or not, and applying Equation 1.5, we have for $|UI| \geq 1$,

$$\prod_{i=0}^{|UI|-1} P\left(UI_{|UI|-i}^{C} \mid \bigcap_{j=0}^{|UI|-(i+1)} UI_j^{C}\right)$$

$$= \prod_{i=0}^{|UI|-1} P\left[k \cdot ln\left(\frac{|UI|-i+noise}{|UI|-(i+1)+noise}\right)\right].$$

In the context of this work, Melamed et al. (2019, p. 2) would imply that when an intruder becomes part of the queue group, this creates a disagreement between the initial queue members and the intruder, generating uncertainty as a result. By conforming to queue group member opinion, a subject of interest within the original queue group reduces his or her uncertainty. Referring to Festinger (1962, p. 95), the terms uncertainty and cognitive dissonance are interchangeable for the given condition. To further make this argument, FeldhamHall and Shenhav (2019, pp. 3–4) relate uncertainty with negative affective reactions. Harmon-Jones et al. (2015, p. 185) using their action-based model suggest that an individual's negative affective state of dissonance is aroused when cognitions with action implications are aroused. Maikovich (2005) uses the radicalization process in terrorist organizations to create a link between cognitive dissonance, low tolerance for uncertainty, and those who resort to violent terrorism. Beasley (2016, pp. 7, 10–11) notes that conditions which involve uncertainty and violation of expectations (which would include social norms) are particularly prone to produce dissonance in an individual. Finally, Randles et al. (2015, p. 706) provide an interpretation of their findings indicating that dissonance may lead to uncertainty. It seems the academic community has been consistently indicating some link between cognitive dissonance and uncertainty since 1962. Additionally, uncertainly reduction begins before the choice is made, or in this case, the choice of reacting or not reacting.

What has been shown so far is that

$$P\left(R_{+|UI|+1}\right) = G_{ES}\left(\overline{es}_{p,m}\right) \cdot P\left(\bigcap_{i=0}^{|UI|} UI_i^C\right)$$

$$= G_{ES}\left(\overline{es}_{p,m}\right) \cdot \prod_{i=0}^{|UI|-1} P\left[k \cdot ln\left(\frac{|UI|-i+noise}{|UI|-(i+1)+noise}\right)\right].$$

For $UI = \left\{\varnothing_0, NR_1, NR_2, \dots, NR_{|N|}\right\}$, $P\left[Z > z\left(|UI|\right)\right]$ for the random variable Z is defined as

$$P\left[Z > z\left(|UI|\right)\right] = \prod_{i=0}^{|UI|-1} P\left[k \cdot ln\left(\frac{|UI|-i+noise}{|UI|-(i+1)+noise}\right)\right]$$

Where the probability function $P\left[Z > z\left(|UI|\right)\right]$ is just the survival function indicating the probability a subject of interest has not yet reacted based on the amount of uncertainty-based information available at the time. For the constraints provided, and the derivation of $P\left[Z > z\left(|UI|\right)\right]$, application of the negative exponential (Efthimiou, 2010, p. 83) for $P\left[Z > z\left(|UI|\right)\right]$ is appropriate. As an example, using $|UI| = 2$ and the function $z\left(|UI|\right)$ when applied to the standard negative exponential,

$$G_{ES}\left(\overline{es}_{p,m}\right) \cdot e^{-z(2)} = G_{ES}\left(\overline{es}_{p,m}\right) \cdot e^{-k \cdot ln\left(\frac{2+noise}{1+noise}\right)} \cdot e^{-k \cdot ln\left(\frac{1+noise}{noise}\right)}$$

$$= G_{ES}\left(\overline{es}_{p,m}\right) \cdot e^{-k \cdot \left[ln\left(\frac{2+noise}{1+noise}\right) + ln\left(\frac{1+noise}{noise}\right)\right]}$$

$$= G_{ES}\left(\overline{es}_{p,m}\right) \cdot e^{-k \cdot \left[ln\left(\frac{2+noise}{noise}\right)\right]}.$$

Constraints may be shown to have been met using the example $G_{ES}\left(\overline{es}_{p,m}\right) \cdot e^{-z(2)}$ and a proof via induction, assuming $0 < noise < \infty$, so that for both $k > 0$ *and* $0 < noise < \infty$,

$$\lim_{|UI| \to \infty} G_{ES}\left(\overline{es}_{p,m}\right) \cdot e^{-k \cdot \left[ln\left(\frac{|UI| + noise}{noise}\right)\right]}$$

$$= 0 \le G_{ES}\left(\overline{es}_{p,m}\right) \cdot e^{-k \cdot \left[ln\left(\frac{|UI| + noise}{noise}\right)\right]} \le G_{ES}\left(\overline{es}_{p,m}\right).$$

Accounting for the independence of certain random variables, this allows derivation of probability of reaction for some specified uncertainty-based information set $UI = \left\{\varnothing_0, NR_1, NR_2, \ldots, NR_{|N|}\right\}, |UI| \ge 0$, and given values $\overline{es}$, p, and m, such that

$$P\left[ES \le \overline{es}_{p,m}, Z > z\left(|UI|\right)\right] = G_{ES}\left(\overline{es}_{p,m}\right) \cdot e^{-z\left(|UI|\right)}$$

$$= \left(1 - e^{-\overline{es}_{p,m}}\right) \cdot e^{-z\left(|UI|\right)}$$

Equation 3.2

With

$$z\left(|UI|\right) = \sum_{i=0}^{|UI|-1} k \cdot ln\left(\frac{|UI| - i + noise}{|UI| - (i+1) + noise}\right) for\ |UI| \ge 1,$$

$$else\ z\left(|UI|\right) = 0.$$

The random variables *ES and Z* are independent if the number of members behind the position of interest and do not depend on $|UI|$. With that, we now have the tools to derive the variables k and *noise* for the function $z\left(|UI|\right)$.

Using data from Table 3.6, we may now evaluate Condition 2 using Equation 3.2. Using $\overline{es}_{L,1} = 0.114$ and $\overline{es}_{H,1} = 0.456$, as derived from Schmitt et al. (1992) and

TABLE 3.6 Condition 2 Empirical Data

Q(P|N|M) = Q(P intruders | N queue members between intruder and position of interest | M queue members behind queue position of interest) = Position Probability of Occurrence				
Position (-1)	Position (-2)	Position (-3)	Position (-4)	Front of Queue
3/24	0/22	0/18	0/3	# React /# Opportunities
1 Intruder (Condition 2)	Notes: 1) Appendix A has k=2 and l_{100}=62.7: theoretical M = 1.4 ≈ documented M_{mean} = 1.2 in the Q(1|1|*) configuration : No extrapolation for "*M*" necessary. 2) The original 1978 summary data and two of the graduate student reports indicate 0/8 at (+4) versus 0/9 per Milgram et al (1986). Milgram et al (1986) takes precedence.			
Position (+1)	Position (+2)	Position (+3)	Position (+4)	Position (+5)
0/24	4/24	0/15	0/9	0/5
Confederate (Graduate Student) Non-reactive Buffer	Q(1|1|3) = 5/24	Q(1|2|2) = 5/15	Q(1|3|1) = 5/9	Q(1|4|0) = 5/5
	Q(1|1|2) = 4/24	Q(1|2|1) = 4/15	Q(1|3|0) = 4/9	Milgram et al (1986)
	Q(1|1|1) = 6/24	Q(1|2|0) = 6/15	supplemented by unpublished data obtained from	
	Q(1|1|0) = 9/24	The Stanley Milgram Papers (MS 1406), Series No. IV, Box No. 100, Folder No. 131		
	Manuscripts and Archives, Yale University Library with copyright permission from the Estate of Alexandra Milgram.			

Condition 1: (+1) of Milgram et al. (1986), we begin by deriving $G_{ES}\left(\overline{es}_{I,1,m}\right)\cdot e^{-z(1)}$ for Condition 2 position (+2). The derivation process is the same as for Conditions 1 and 4 at position (+1) using the Law of Total Probability while incorporating Equation 3.2 with $UI = \{NR_{+1}\} \Rightarrow |UI| = 1 \; and \; P\left(UI = \{NR_{+1}\}\right) = 1$. It is worth repeating due to lack of experimental raw data, homogeneity with regard to m of high (H) and low (L) unit stimulus sensation magnitude queues must be assumed.

$$P\left[ES \le \overline{es}_{I,1,m}, \overline{es}_H, Q(1|1|0)\right] \cdot P\left[Z > z\left(|UI|\right)\right]$$

$$= \left(1 - e^{-.456}\right) \cdot \left(\frac{19\,\overline{es}_H\,Trials}{24\,Trials}\right) \cdot \left(\frac{9}{24}\right) \cdot e^{-z(1)}$$

$$= 0.109 \cdot e^{-z(1)}$$

$$P\left[ES \le \overline{es}_{I,1,m}, \overline{es}_H, Q(1|1|1)\right] \cdot P\left[Z > z\left(|UI|\right)\right]$$

$$= \left(1 - e^{-0.456 \cdot ln\left(e \cdot \frac{1+1.147}{1.147}\right)}\right) \cdot \left(\frac{19}{24}\right) \cdot \left(\frac{6}{24}\right) \cdot e^{-z(1)}$$

$$= 0.104 \cdot e^{-z(1)}$$

$$P\left[ES \le \overline{es}_{I,1,m}, \overline{es}_H, Q(1|1|2)\right] \cdot P\left[Z > z\left(|UI|\right)\right]$$

$$= \left(1 - e^{-0.456 \cdot ln\left(e \cdot \frac{2+1.147}{1.147}\right)}\right) \cdot \left(\frac{19}{24}\right) \cdot \left(\frac{4}{24}\right) \cdot e^{-z(1)}$$

$$= 0.079 \cdot e^{-z(1)}$$

$$P\left[ES \le \overline{es}_{I,1,m}, \overline{es}_H, Q(1|1|3)\right] \cdot P\left[Z > z\left(|UI|\right)\right]$$

$$= \left(1 - e^{-0.456 \cdot ln\left(e \cdot \frac{3+1.147}{1.147}\right)}\right) \cdot \left(\frac{19}{24}\right) \cdot \left(\frac{5}{24}\right) \cdot e^{-z(1)}$$

$$= 0.107 \cdot e^{-z(1)}$$

$$P\left[ES \le \overline{es}_{I,1,m}, \overline{es}_L, Q(1|1|0)\right] \cdot P\left[Z > z\left(|UI|\right)\right]$$

$$= \left(1 - e^{-0.114}\right) \cdot \left(\frac{5\,\overline{es}_L\,Trials}{24\,Trials}\right) \cdot \left(\frac{9}{24}\right) \cdot e^{-z(1)}$$

$$= 0.008 \cdot e^{-z(1)}$$

$$P\left[ES \leq \overline{es}_{I,1,m}, \overline{es}_L, Q(1|1|1)\right] \cdot P\left[Z > z\left(|UI|\right)\right]$$

$$= \left(1 - e^{-0.114 \cdot ln\left(e \cdot \frac{1+1.147}{1.147}\right)}\right) \cdot \left(\frac{5}{24}\right) \cdot \left(\frac{6}{24}\right) \cdot e^{-z(1)}$$

$$= 0.009 \cdot e^{-z(1)}$$

$$P\left[ES \leq \overline{es}_{I,1,m}, \overline{es}_L, Q(1|1|2)\right] \cdot P\left[Z > z\left(|UI|\right)\right]$$

$$= \left(1 - e^{-0.114 \cdot ln\left(e \cdot \frac{2+1.147}{1.147}\right)}\right) \cdot \left(\frac{5}{24}\right) \cdot \left(\frac{4}{24}\right) \cdot e^{-z(1)}$$

$$= 0.007 \cdot e^{-z(1)}$$

$$P\left[ES \leq \overline{es}_{I,1,m}, \overline{es}_L, Q(1|1|3)\right] \cdot P\left[Z > z\left(|UI|\right)\right]$$

$$= \left(1 - e^{-0.114 \cdot ln\left(e \cdot \frac{3+1.147}{1.147}\right)}\right) \cdot \left(\frac{5}{24}\right) \cdot \left(\frac{5}{24}\right) \cdot e^{-z(1)}$$

$$= 0.010 \cdot e^{-z(1)}$$

Accounting for round-off error in the aforementioned and as before assuming $P\left[\overline{es}_I \mid Q(1|0|m)\right] = P\left[\overline{es}_I\right]$,

$$P\left(R_{+|UI|+1}\right) = P\left(R_{+2}\right)$$

$$= \sum_{I \in \{H,L\}} \sum_{m \in M} P\left[ES \leq \overline{es}_{I,1,m}, \overline{es}_I, Q(1|1|m)\right] \cdot P\left[Z > z\left(|UI|\right)\right]$$

$$= \sum_{I \in \{H,L\}} \sum_{m \in M} G_{ES}\left(\overline{es}_{I,1,\{NR_1\},m}\right) \cdot P\left(\overline{es}_I\right) \cdot P\left[Q(1|1|m)\right] \cdot e^{-z(1)}$$

$$= 0.433 \cdot e^{-z(1)}$$

$$= \frac{4}{24} = 0.167 \left(empirical\ result\right).$$

$$e^{-z(1)} = \frac{0.167}{0.433} = 0.386, or\ z(1) = -ln(0.386) = 0.952.$$

TABLE 3.7 Condition 5 Empirical Data

$Q(P\|N\|M) = Q(P$ intruders | N queue members between intruder and position of interest | M queue members behind queue position of interest) = Position Probability of Occurrence				
Position (-1)	Position (-2)	Position (-3)	Position (-4)	Front of Queue
2/20	0/18	0/13	0/7	# React /# Opportunities
2 Intruders (Condition 5)	**Notes:** 1) Appendix A has k=10 and l_{100}=61.1: theoretical M = 1.1 ≈ documented M_{mean} = 1.0 in the $Q(2\|1\|^*)$ configuration: No extrapolation for "*M*" necessary. 2) Original 1978 summary data and two graduate reports indicate 5/20 for position (+2) versus 4/20 as reported in Milgram et al (1986). Milgram et al (1986) takes precedence.			
Position (+1)	Position (+2)	Position (+3)	Position (+4)	Position (+5)
0/20	4/20	0/15	0/4	0/0
Confederate (Graduate Student) Non-reactive Buffer	$Q(2\|1\|3) = 0/20$ $Q(2\|1\|2) = 4/20$ $Q(2\|1\|1) = 11/20$ $Q(2\|1\|0) = 5/20$	$Q(2\|2\|2) = 0/15$ $Q(2\|2\|1) = 4/15$ $Q(2\|2\|0) = 11/15$	$Q(2\|3\|1) = 0/4$ $Q(2\|3\|0) = 4/4$	No Opportunities Milgram et al (1986) supplemented by unpublished data obtained from The Stanley Milgram Papers (MS 1406), Series No. IV, Box No. 100, Folder No. 131 Manuscripts and Archives, Yale University Library with copyright permission from the Estate of Alexandra Milgram.

With $z(1)$ derived, we may now evaluate Condition 5, which also has a single buffer as Condition 2, and check for repeatability using a queue experiencing two simultaneous intruders.

Applying $z(1) = 0.952$ as derived, in the same manner as Condition 2, it is found by spreadsheet calculations that $\overline{es}_{H,2} = 0.780$ and $\overline{es}_{L,2} = 0.195$ when the total probability of reaction equals $\frac{4}{20}$ (*i.e.,* 0.200) for Condition 5: (+2), see Table 3.7. The sample mean of $\overline{es}_{H,2} = 0.780$ is a feasible solution with Conditions 1 and 2 as per Corollary 2.3. We can now move to Condition 6 and get an approximate value for z(2). If $z(1)$ and $z(2)$ meet Corollary 2.3 requirements, then Theorem 2.5 may be applied to calculate a noise value related to member interpretation of intermediate member reactions. Table 3.8 summarizes the necessary Condition 6 reaction data as a function of position.

The same approach as Condition 2 and 5 is used for Condition 6: (+3) but using $\overline{es}_{H,2} = 0.780$ (note: there are no low sensation magnitude queues in this condition). Given $UI = \{NR_{+1}, NR_{+2}\}$ and that $P(UI) = 1$, the result is,

$$P(R_{+3}) = \sum_{I \in \{H\}} \sum_{m \in M} P\left[ES \leq \overline{es}_{I,2,m}, \overline{es}_I, Q(2\|2\|m) \right]$$

$$\cdot P\left[Z > z(|UI|) \right] = \sum_{I \in \{H\}} \sum_{m \in M} G_{ES}\left(\overline{es}_{I,2,m} \right)$$

$$\cdot P(\overline{es}_I) \cdot P\left[Q(2\|2\|m) \right] \cdot e^{-z(2)} = 0.731 \cdot e^{-z(2)}$$

$$= \frac{3}{20} = 0.150 \,(empirical\ result).$$

$$As\ before,\ e^{-z(2)} = \frac{0.150}{0.731} = 0.205, or\ z(2)$$

$$= -ln(0.205) = 1.585.$$

TABLE 3.8 Condition 6 Empirical Data

$Q(P|N|M) = Q(P$ intruders $\|$ N queue members between intruder and position of interest $\|$ M queue members behind queue position of interest) = Position Probability of Occurrence				
Position (-1) ⬇	⬅ Position (-2)	⬅ Position (-3)	⬅ Position (-4)	⬅ Front of Queue
2/20	0/18	0/18	0/13	# React /# Opportunities
👥 **2 Intruders (Condition 6)** ⬇	**Notes**: 1) Appendix A has k=10 and l_{100}=76.3: theoretical M = 2.1 slightly > documented M_{mean} = 1.3 in the Q(2|2|*) configuration: No extrapolation for "*M*" necessary. 2) Original 1978 summary data and two graduate reports indicate 2/16 for position (+4) versus 2/17 as reported in Milgram et al (1986). Milgram et al (1986) takes precedence.			
Position (+1) ⬇	➡ Position (+2)	➡ Position (+3)	➡ Position (+4)	➡ Position (+5)
0/20	0/20	3/20	2/17	0/10
Confederate (Graduate Student) Non-reactive Buffer	**Confederate (Graduate Student) Non-reactive Buffer**	$Q(2|2|2) = 10/20$ $Q(2|2|1) = 7/20$ $Q(2|2|0) = 3/20$	$Q(2|3|1) = 10/17$ $Q(2|3|0) = 7/17$ supplemented by unpublished data obtained from The Stanley Milgram Papers (MS 1406), Series No. IV, Box No. 100, Folder No. 131 Manuscripts and Archives, Yale University Library with copyright permission from the Estate of Alexandra Milgram.	$Q(2|4|0) = 10/10$ Milgram et al (1986)

It may also be observed that Corollary 2.3 applies since $1 < \frac{z(2)}{z(1)} = \frac{1.585}{0.952} < 2$.

As expected, a logarithmic approach to modeling presents itself as predicted from the derivation of $z(|N|)$ by noting that for two confederate nonreactive buffers,

$$z(2) = 1.585 \cong z(1) + z(1) \cdot ln(2) = 0.952 \cdot ln(e \cdot 2) = 1.612.$$

Hence, a feasible value exists for a noise term from the data related to a member's ability to observe and interpret the non-reactions of other group members closer to the dissonant-causing event. In this case, non-reaction by a queue member immediately behind the intrusion point provides information to other queue members. Non-reaction implies to others that the intrusion did not appear to be a significant deviation of group norms and therefore is likely to be of low-sensation magnitude. The member immediately following may use this information to reduce his or her uncertainty as to what occurred and thereby determine whether to react or not react.

In the example of two nonreactive queue members immediately behind the intrusion point, the amount of uncertainty as measured in probability space by queue members in subsequent positions is about one-fifth (i.e., $e^{-z(2)} = 0.205$) of the initial amount. The uncertainty-based information noise variable N_{UI} may be noise interfering with obtaining uncertainty-based information, or it may be a constant. If noise increases, the ability of the member of interest to interpret and make use of information through observation of pertinent members decreases. Described in this way, it would seem N_{UI} and N_{UA} should be positively correlated, given members in attractive (i.e., cohesive) groups would be better able to interpret one another's reactions toward any deviation of group social norms.

Given the values for $z(1)$ and $z(2)$ meet Corollary 2.3 requirements, Theorem 2.4 and Theorem 2.5 may be applied to find $\hat{N}_{UI}$ and $\overline{es}(\hat{N}_{UI})$, respectively. The value of $\hat{N}_{UI}$ supporting,

$$
\left[0.952 - \frac{1.585 \cdot ln\left(\dfrac{1+\widehat{N}_{UI}}{\widehat{N}_{UI}}\right)}{ln\left(\dfrac{2+\widehat{N}_{UI}}{\widehat{N}_{UI}}\right)} \right] + \left[1.585 - \frac{0.952 \cdot ln\left(\dfrac{2+\widehat{N}_{UI}}{\widehat{N}_{UI}}\right)}{ln\left(\dfrac{1+\widehat{N}_{UI}}{\widehat{N}_{UI}}\right)} \right]
$$
$$
= 0
$$

is $\widehat{N}_{UI} = 1.553$. Using $\widehat{N}_{UI} = 1.553$ with Theorem 2.4 results in

$$
\overline{es}\left(\widehat{N}_{UI}\right) = \frac{1}{2} \cdot \left[\frac{0.952}{ln\left(\dfrac{1+\widehat{N}_{UI}}{\widehat{N}_{UI}}\right)} + \frac{1.585}{ln\left(\dfrac{2+\widehat{N}_{UI}}{\widehat{N}_{UI}}\right)} \right] = 1.195.
$$

From this, we have

$$
\overline{z}(1) = 1.915 \cdot ln\left(\frac{1+1.553}{1.553}\right) = 0.952;
$$

$$
\overline{z}(2) = 1.915 \cdot ln\left(\frac{2+1.553}{1.553}\right) = 1.585.
$$

Stated in probability space, $e^{-\overline{z}(1)} = e^{-0.952} = 0.386,\ or\ 1 - e^{-\overline{z}(1)} = 0.614$.

The probability 0.614 is extremely close to a very familiar value. The value 0.618 is the inverse of the Golden Ratio. In operations research, the value 0.618 is used in the Golden Section search to optimize speed of finding the maximum value of a unimodal function (Winston, 1987, pp. 545–549) with minimal computational effort. Referring to Tanner et al. (1967, p. 352), based on a least squares fit of the data, probability of the observers identifying the sound as the target sound approaches the inverse Golden Ratio as the probability of the target sound being introduced approaches 1. Benjafield and Adams-Weber (1976) provide a hypothesis leading to even more interesting and potentially pertinent results. Let positive events be associated with a non-reaction indicating acceptable behavior by group members, and let negative events be associated with a reaction to indicate unacceptable behavior under those same norms. Benjafield and Adams-Weber (1976, pp. 13–14) hypothesize that socially deviant behaviors become most noticeable when compared to the acceptable background of non-deviant behaviors when the proportion of negative events (social deviation) divided by the proportion of positive events (group-accepted behavior) tends to the Golden Section value of 0.618.

In summary, for $|UI|$ nonreactive buffers between the intruder(s) and the $(+|UI| + 1)$ position of interest, $e^{-z(|UI|)}$ may be expressed for the New York queue as

TABLE 3.9 Condition 3 Empirical Data

$Q(P|N|M) = Q(P \text{ intruders} \mid N \text{ queue members between intruder and position of interest} \mid$ $M \text{ queue members behind queue position of interest}) = \text{Position Probability of Occurrence}$				
Position (-1)	⬅ Position (-2)	⬅ Position (-3)	⬅ Position (-4)	⬅ Front of Queue
1/20	0/18	0/17	0/9	# React /# Opportunities
⬆ 1 Intruder (Condition 3) ⬇	Notes: 1) Appendix A has k=10 and l_{100}=74.7: theoretical M = 2 > documented M_{mean} = 1.4 in the Q(1|2|*) configuration: No difference though since current Q(1|2|2) = theoretical Q(1|2|2): No extrapolation for "M" necessary.			
Position (+1)	➡ Position (+2)	➡ Position (+3)	➡ Position (+4)	➡ Position (+5)
0/20	0/20	0/20	0/18	0/9
Confederate (Graduate Student) Non-reactive Buffer	Confederate (Graduate Student) Non-reactive Buffer	$Q(1|2|2) = 9/20$ $Q(1|2|1) = 9/20$ $Q(1|2|0) = 2/20$	$Q(1|3|1) = 9/18$ $Q(1|3|0) = 9/18$ supplemented by unpublished data obtained from	$Q(1|4|0) = 9/9$ Milgram et al (1986)
		The Stanley Milgram Papers (MS 1406), Series No. IV, Box No. 100, Folder No. 131		
		Manuscripts and Archives, Yale University Library with copyright permission from the Estate of Alexandra Milgram.		

$$e^{-\overline{z}(|UI|)} = e^{-1.915 \cdot ln\left(\frac{|UI|+1.553}{1.553}\right)}$$

Equation 3.3

$$\text{where all elements in } UI \text{ are nonreacting.}$$

With Equation 3.3 available, Condition 3 data shown in Table 3.9 may now be analyzed.

Using $\overline{es}_{H,1} = 0.456$ and $\overline{es}_{L,1} = 0.114$ from Condition 1, and applying the same process as demonstrated for Conditions 1 and 6 while using values from Table 3.9, we obtain,

$$P(R_{+3}) = \sum_{I=\{H,L\}} \sum_{m \in M} P\left[ES \leq \overline{es}_{I,1,m}, \overline{es}_I, Q(1|2|m)\right] \cdot P\left[Z > z(|UI|)\right]$$

$$= \sum_{I=\{H,L\}} \sum_{m \in M} G_{ES}\left(\overline{es}_{I,1,m}\right) \cdot P\left[\overline{es}_I\right] \cdot P\left[Q(1|2|m)\right] \cdot e^{-\overline{z}(2)}$$

$$= 0.063.$$

The result of 0.063 for Condition 3: (+3) is well within the 95 percent credibility interval of (0.0, 0.133). We now have all the mathematical tools, when combined with Equation 3.3, to analyze empirical data for Conditions 1 and 4 at the (+1) positions and for Conditions 2, 3, 5, and 6 at the positions just after their respective nonreactive buffers. To solve for the remaining Conditions and positions requires addressing those intermediate members who react between the intruder(s) and position of interest.

3.3.4 Impact of Reacting Members: Deriving y(|UI|) Using Condition 4: (+2)

Experiments and data collection are difficult, and the intrusion experiment by Stanley Milgram and his graduate students in 1978 was even more so. Though the experiment and the data are nothing less than outstanding, there are always lessons learned that should be documented for others to build on. It is also important to understand the data. To this end, in his 1978 student report, graduate student David Niemeroff wrote the following (Stanley Milgram Papers (MS1406), with copyright permission from the Estate of Alexandra Milgram):

We can observe with 2 intruders and no buffers nearly all the lines objected. Interestingly, in this condition the only two lines that did not object were the two information booth lines at Grand Central station.

This clearly refers to Condition 4 since there are two intruders and no buffers. David Niemeroff further notes that these two lines were the fastest and also the most amorphous, alternating their topology every minute or so between that of a straight line or a cluster as is sometimes seen at a bus stop (Stanley Milgram Papers, MS 1406, Box 101, Folder 136, pp. 8–9). This indicates the ambiguity of queue position status. It also indicates potential confusion of roles and responsibilities for maintaining group boundaries separating those in queue from those not in queue. The importance is that being amorphous leads to position uncertainty, social uncertainty (embarrassment if wrong to react), and greater uncertainty of attraction. It is suggested here that the information booth was not representative when viewed within the context of all the other queue locations detailed in Table 3.3.

In 1978, when preparing for the experiment, graduate student Joyce Wackenhut defined the queue in her planning notes as a clearly defined social structure. In his translation of Parsons (1951), Mann (1969, p. 349) provides the properties of a social system of which he states the queue is an embryonic representative. He states that three properties of any social system are: (1) two or more members occupying differentiated statuses or positions and performing differentiated roles; (2) some organized pattern governing the relationships of the members, describing their rights and obligations with respect to one another; and (3) some set of common norms and values, together with various types of shared cultural objects and symbols. As a lesson learned, queues like the information booth queue as described by David Niemeroff are not well-differentiated or defined social systems and should be avoided for queue analysis.

Condition 4: (+1) has the most reactions (20 out of 23) of any (+1) position for the six conditions. This makes it an excellent candidate to derive $y(1)$ using Condition 4: (+2) results by replacing $z(1)$ for $UI = \{NR_{+1}\}$ with $y(1)$ for $UI = \{R_{+1}\}$ in Equation 3.2. Given David Niemeroff's report, it is apparent there are 21 observations of interest here, and of those 21, 20 had reactions at the (+1) position. Of those 21 observations, there were 10 reactions at the (+2) position. With that simple observation, we can guess up front that $e^{-\bar{y}(1)}$ will be somewhere around 0.5. Ultimately, once $e^{-\bar{y}(n)}$ is obtained in a mathematically traceable manner, all remaining Conditions at positions beyond (+1) may be theoretically calculated using Equation 3.2 where the set UI now may contain any ranked combination of reacting and non-reacting members.

Using the Law of Total Probability with Equation 3.2, while now accounting for the possibility of a reactive or nonreactive intermediate queue member, and assuming as before, based on lack of available information that $P\left[\overline{es}_I \mid Q(2|1|m)\right] = P\left(\overline{es}_I\right)$, results in,

$$P(R_{+2}) = \sum_{r \in \{R_{+1}, NR_{+1}\}} \sum_{I \in \{H,L\}} \sum_{m \in M} G_{ES}\left(\overline{es}_{I,2,m}\right)$$

$$\cdot P\left(\overline{es}_I\right) \cdot P\left[Q(2|1|m)\right] \cdot H_{+2}(r) \cdot P(r),$$

$$\text{defining } H_{+2}(r) \cdot P(r) = e^{-\overline{y}(1)} \cdot G\left(\overline{es}_{I,2,m+1}\right)$$

$$\text{if } r = R_{+1} \text{ else } H_{+2}(r) \cdot P(r)$$

$$= e^{-\overline{z}(1)}\left[1 - G\left(\overline{es}_{I,2,m+1}\right)\right], \text{if } r = NR_{+1}.$$

Knowing $\overline{es}_{H,2} = 0.780$, but for this situation using $\overline{es}_{H,2} = 1.005$ as derived for Condition 4: (+1) to solve for $e^{-y(1)}$ under the same conditions, we may now solve for $e^{-y(1)}$ in the following manner using Condition 4: (+2):

$$\sum_{r \in \{R_{+1}, NR_{+1}\}} P\left[ES \le \overline{es}_{I,2,m}, \overline{es}_H, Q(2|1|0)\right] \cdot H_{+2}(r) \cdot P(r)$$

$$= \left(1 - e^{-1.005}\right) \cdot \left(\frac{19 \, \overline{es}_H \, Trials}{23 \, Trials}\right) \cdot \left(\frac{1}{23}\right)$$

$$\cdot \left[e^{-\overline{z}(1)} \cdot \left[1 - G\left(\overline{es}_{H,2,m}\right)\right] + e^{-\overline{y}(1)} \cdot G\left(\overline{es}_{H,2,m}\right)\right]$$

$$= 0.0228 \cdot \left[0.386 \cdot e^{-1.005 \cdot ln\left(e \cdot \frac{1+1.147}{1.147}\right)} + e^{-y(1)}\right.$$

$$\left. \cdot \left(1 - e^{-1.005 \cdot ln\left(e \cdot \frac{1+1.147}{1.147}\right)}\right)\right] = 0.0017 + 0.0183 \cdot e^{-y(1)}$$

$$\sum_{r \in \{R_{+1}, NR_{+1}\}} P\left[ES \le \overline{es}_{I,2,m}, \overline{es}_H, Q(2|1|1)\right] \cdot H_{+2}(r) \cdot P(r)$$

$$= \left(1 - e^{-1.005 \cdot ln\left(e \cdot \frac{1+1.147}{1.147}\right)}\right) \cdot \left(\frac{19}{23}\right) \cdot \left(\frac{2}{23}\right)$$

$$\cdot \left[e^{-\overline{z}(1)} \cdot \left[1 - G\left(\overline{es}_{H,2,m}\right)\right] + e^{-y(1)} \cdot G\left(\overline{es}_{H,2,m}\right)\right]$$

$$= 0.0578 \cdot \left[0.386 \cdot e^{-1.005 \cdot ln\left(e^{\frac{2+1.147}{1.147}}\right)} + e^{-y(1)} \cdot \left(1 - e^{-1.005 \cdot ln\left(e^{\frac{2+1.147}{1.147}}\right)}\right)\right]$$

$$= 0.003 + 0.050 \cdot e^{-y(1)}$$

$$\sum_{r \in \{R_{+1}, NR_{+1}\}} P\left[ES \leq \overline{es}_{I,2,m}, \overline{es}_H, Q(2|1|2)\right] \cdot H_{+2}(r) \cdot P(r)$$

$$= \left(1 - e^{-1.005 \cdot ln\left(e \cdot \frac{2+1.147}{1.147}\right)}\right) \cdot \left(\frac{19}{23}\right) \cdot \left(\frac{3}{23}\right)$$

$$\cdot \left[e^{-\overline{z}(1)} \cdot \left[1 - G\left(\overline{es}_{H,2,m}\right)\right] + e^{-\overline{y}(1)} \cdot G\left(\overline{es}_{H,2,m}\right)\right]$$

$$= 0.0934 \cdot \left[0.386 \cdot e^{-1.005 \cdot ln\left(e \cdot \frac{3+1.147}{1.147}\right)} + e^{-y(1)} \cdot \left(1 - e^{-1.005 \cdot ln\left(e \cdot \frac{3+1.147}{1.147}\right)}\right)\right]$$

$$= 0.0036 + 0.0840 \cdot e^{-y(1)}$$

$$\sum_{r \in \{R_{+1}, NR_{+1}\}} P\left[ES \leq \overline{es}_{I,2,m}, \overline{es}_H, Q(2|1|10.3)\right] \cdot H_{+2}(r) \cdot P(r)$$

$$= \left(1 - e^{-1.005 \cdot ln\left(e \cdot \frac{10.3+1.147}{1.147}\right)}\right) \cdot \left(\frac{19}{23}\right) \cdot \left(\frac{17}{23}\right)$$

$$\cdot \left[e^{-\overline{z}(1)} \cdot \left[1 - G\left(\overline{es}_{H,2,m}\right)\right] + e^{-\overline{y}(1)} \cdot G\left(\overline{es}_{H,2,m}\right)\right]$$

$$= 0.588 \cdot \left[0.386 \cdot e^{-1.005 \cdot ln\left(e \cdot \frac{11.3+1.147}{1.147}\right)} + e^{-y(1)} \cdot \left(1 - e^{-1.005 \cdot ln\left(e \cdot \frac{11.3+1.147}{1.147}\right)}\right)\right]$$

$$= 0.0076 + 0.5688 \cdot e^{-y(1)}$$

$$\sum_{UI \in \{R_{+1}, NR_{+1}\}} P\left[ES \leq \overline{es}_{I,2,m}, \overline{es}_L, Q(2|1|0)\right] \cdot H_{+2}(r) \cdot P(r)$$

$$= \left(1 - e^{-0.251}\right) \cdot \left(\frac{4\overline{es}_L \text{ Trials}}{23 \text{ Trials}}\right) \cdot \left(\frac{1}{23}\right) \cdot$$

$$\cdot \left[e^{-\overline{z}(1)} \cdot \left[1 - G\left(\overline{es}_{H,2,m}\right)\right] + e^{-\overline{y}(1)} \cdot G\left(\overline{es}_{H,2,m}\right)\right]$$

$$= 0.0017 \cdot \left[0.386 \cdot e^{-0.251 \cdot ln\left(e \cdot \frac{1+1.147}{1.147}\right)} + e^{-y(1)} \cdot \left(1 - e^{-0.251 \cdot ln\left(e \cdot \frac{1+1.147}{1.147}\right)}\right)\right]$$

$$= 0.0004 + 0.0006 \cdot e^{-y(1)}$$

$$\sum_{UI \in \{R_{+1}, NR_{+1}\}} P\left[ES \leq \overline{es}_{I,2,m}, \overline{es}_L, Q(2|1|1)\right] \cdot H_{+2}(r) \cdot P(r)$$

$$= \left(1 - e^{-0.251 \cdot ln\left(e \cdot \frac{1+1.147}{1.147}\right)}\right) \cdot \left(\frac{4}{23\,Trials}\right) \cdot \left(\frac{2}{23}\right)$$

$$\cdot \left[e^{-\overline{z}(1)} \cdot \left[1 - G\left(\overline{es}_{H,2,m}\right)\right] + e^{-\overline{y}(1)} \cdot G\left(\overline{es}_{H,2,m}\right)\right]$$

$$= 0.0051 \cdot \left[0.386 \cdot e^{-0.251 \cdot ln\left(e \cdot \frac{2+1.147}{1.147}\right)} + e^{-y(1)}\right.$$

$$\left. \cdot \left(1 - e^{-0.251 \cdot ln\left(e \cdot \frac{2+1.147}{1.147}\right)}\right)\right]$$

$$= 0.0012 + 0.0020 \cdot e^{-y(1)}$$

$$\sum_{UI \in \{R_{+1}, NR_{+1}\}} P\left[ES \leq \overline{es}_{I,2,m}, \overline{es}_L, Q(2|1|2)\right] \cdot H_{+2}(r) \cdot P(r)$$

$$= \left(1 - e^{-0.251 \cdot ln\left(e \cdot \frac{2+1.147}{1.147}\right)}\right) \cdot \left(\frac{4}{23\,Trials}\right) \cdot \left(\frac{3}{23}\right)$$

$$\cdot \left[e^{-\overline{z}(1)} \cdot \left[1 - G\left(\overline{es}_{H,2,m}\right)\right] + e^{-\overline{y}(1)} \cdot G\left(\overline{es}_{H,2,m}\right)\right]$$

$$= 0.009 \cdot \left[0.386 \cdot e^{-0.251 \cdot ln\left(e \cdot \frac{3+1.147}{1.147}\right)} + e^{-y(1)}\right.$$

$$\left. \cdot \left(1 - e^{-0.251 \cdot ln\left(e \cdot \frac{3+1.147}{1.147}\right)}\right)\right]$$

$$= 0.002 + 0.0039 \cdot e^{-y(1)}$$

$$\sum_{UI \in \{R_{+1}, NR_{+1}\}} P\left[ES \le \overline{es}_{I,2,m}, \overline{es}_L, Q(2|1|10.3)\right] \cdot H_{+2}(r) \cdot P(r)$$

$$= \left(1 - e^{-0.251 \cdot ln\left(e \cdot \frac{10.3+1.147}{1.147}\right)}\right) \cdot \left(\frac{4}{23\,Trials}\right) \cdot \left(\frac{17}{23}\right)$$

$$\cdot \left[e^{-\overline{z}(1)} \cdot \left[1 - G\left(\overline{es}_{H,2,m}\right)\right] + e^{-\overline{y}(1)} \cdot G\left(\overline{es}_{H,2,m}\right)\right] = 0.0724$$

$$\cdot \left[0.386 \cdot e^{-0.251 \cdot ln\left(e \cdot \frac{11.3+1.147}{1.147}\right)} + e^{-y(1)} \cdot \left(1 - e^{-0.251 \cdot ln\left(e \cdot \frac{11.3+1.147}{1.147}\right)}\right)\right]$$

$$= 0.0120 + 0.0414 \cdot e^{-y(1)}$$

Now add the terms just calculated and equate the sum to the empirical result of 0.435 representing the Condition 4: (+2) empirical probability of reaction to obtain,

$$0.0314 + 0.7693 \cdot e^{-y(1)} = 0.435, \text{ or } e^{-y(1)} = 0.5246$$

Which solves to $y(1) = 0.645$. For now, making the rather large assumption that noise, assuming that is what it is, caused by uncertainty of interpretation is the same as that for $z(1)$ given both are in the same social situation, we obtain,

$$\overline{y}(|UI|) = \frac{0.645}{ln\left(\frac{1+1.553}{1.553}\right)} \cdot ln\left(\frac{|UI|+1.553}{1.553}\right)$$

$$= 1.298 \cdot ln\left(\frac{|UI|+1.553}{1.553}\right), \text{ for } UI$$

$$= \{R_n, R_{n-1}, \ldots, R_1, \varnothing\}, n \in \mathbb{Z}^+.$$

Equation 3.4

Like the value obtained for $e^{-\overline{z}(1)}$, the value obtained for $e^{-\overline{y}(1)}$ has potential significance as well. In the queue, the binary information obtained is either a reaction directed at the intruder or non-reaction. In information theory, the amount of uncertainty-based information obtained is maximized (highest entropy) when the probability of either binary event is 0.5 (Stone, 2015, pp. 33–38). Interpreting this, it may be that as a species, we minimize effort to confirm that everything is normal socially but maximize information intake when a confirmed social deviation event occurs. It seems that a significant amount of information theory is woven into this process, and it is felt that those who have expertise in this area would likely find further investigation fruitful. It is also possible, given the values derived for $z(1)$ and $y(1)$ and their potential interpretation, that $\widehat{N}_{UI}$ may be a constant, and a very interesting one if with more data it is found that

$$\frac{1+\widehat{N}_{UI}}{\widehat{N}_{UI}} \to 1.618 \; since \; \frac{1+1.553}{1.553} = 1.644 \; is \; rather \; close \; to$$

the Goldern Ratio already.

As has been observed in the intrusion data from this chapter, some of the intermediate queue members react, while others do not. This fact requires a generalization which may be found by considering $n = |UI| \in \mathbb{Z}^0$ and the set $UI = \{\varnothing_0, UI_1, UI_2, \ldots, UI_n\}$ where each $UI_{i+1} \in \{R_{i+1}, NR_{i+1}\}$ with $i \in \mathbb{Z}^0$, $0 \leq i \leq |UI|-1$ so that

$$H_{|UI|+1}(UI) = \prod_{i=0}^{|UI|-1} p\left[k\left(UI_{i+1}\right) \cdot ln\left(\frac{|UI|-i+noise}{|UI|-(i+1)+noise} \right) \right],$$

recalling $|\varnothing| = 0$.

This is equivalent to,

$$H_{|UI|+1}(UI) = e^{-\sum_{i=o}^{|UI|-1} k(UI_{i+1}) \cdot ln\left(\frac{|UI|-i+noise}{|UI|-(i+1)+noise} \right)}$$

Equation 3.5

for $|UI| \geq 1$, *and* $H_{+1}(\varnothing) = 1$, *with* $k\left(R_{i+1}\right) = 1.298$
and $k\left(NR_{i+1}\right) = 1.915$.

Using Equation 3.5, $UI = \{\varnothing_0, UI_1, UI_2, \ldots, UI_n\}$, $UI \in \{R_{+i}, NR_{+i}\}$, and m members behind the subject of interest having $|UI|$ intermediate queue members between him or her and the intruder(s), define,

$$P\left(R_{+|UI|+1}, UI\right) = G\left(\overline{es}_{p,m}\right) \cdot H_{|UI|+1}(UI)$$

$$\cdot P\left(UI_{|UI|}, UI_{|UI|-1}, UI_{|UI|-2}, \ldots, UI_1\right),$$

so that for all possible combinations of UI,

$$P\left(R_{+|UI|+1}\right) = \sum\left[G\left(\overline{es}_{p,m}\right)\cdot H_{|UI|+1}\left(UI\right)\right]$$
$$\cdot P\left(UI_{|UI|},UI_{|UI|-1},UI_{|UI|-2},\ldots,UI_1,\varnothing\right)$$

$$using\left[G\left(\overline{es}_{p,m}\right)\cdot H_{|UI-\{R_{|UI|}\}|+1}\left(UI-\{R_{|UI|}\}\right)\right]$$

$$for\ P\left(R_{|UI|}\mid UI_{|UI|-1},UI_{|UI|-2},\ldots,UI_1,\varnothing\right),and$$

$$\left[1-G\left(\overline{es}_{p,m}\right)\cdot H_{|UI-\{NR_{|UI|}\}|+1}\left(UI-\{NR_{|UI|}\}\right)\right]$$

$$for\ P\left(NR_{|UI|}\mid UI_{|UI|-1},UI_{|UI|-2},\ldots,UI_1,\varnothing\right)$$

and so on until all the elements of *UI* have been addressed.

To provide an example of this process, consider the situation where two queue members stand between the intruder(s) and member of interest at queue position (+3). Let the number of members behind the (+3) position be given as m. Given $UI = \{UI_{+2},UI_{+1},\varnothing\}$, the probability of the member of interest reacting, $P\left(R_{+3}\right)$, given all possible combinations of *UI* is found by calculating,

$$P\left(R_{+3}\right) = \sum\left[G\left(\overline{es}_{p,m}\right)\cdot H_{|UI|+1}\left(UI\right)\right]\cdot P\left(UI\right)$$
$$= G\left(\overline{es}_{p,m}\right)\cdot H_{+3}\left(R_{+2},NR_{+1},\varnothing\right)\cdot\left[G\left(\overline{es}_{p,m+1}\right)\cdot H_{+2}\left(NR_{+1},\varnothing\right)\right]$$
$$\cdot\left[1-G\left(\overline{es}_{p,m+2}\right)\cdot H_{+1}\left(\varnothing\right)\right]+G\left(\overline{es}_{p,m}\right)H_{+3}\left(R_{+2},R_{+1},\varnothing\right)$$
$$\cdot\left[G\left(\overline{es}_{p,m+1}\right)\cdot H_{+2}\left(R_{+1},\varnothing\right)\right]\cdot\left[G\left(\overline{es}_{p,m+2}\right)\cdot H_{+1}\left(\varnothing\right)\right]+G\left(\overline{es}_{p,m}\right)$$
$$\cdot H_{+3}\left(NR_{+2},NR_{+1},\varnothing\right)\cdot\left[1-G\left(\overline{es}_{p,m+1}\right)\cdot H_{+2}\left(NR_{+1},\varnothing\right)\right]$$
$$\cdot\left[1-G\left(\overline{es}_{p,m+2}\right)\cdot H_{+1}\left(\varnothing\right)\right]+G\left(\overline{es}_{p,m}\right)\cdot H_{+3}\left(NR_{+2},R_{+1},\varnothing\right)$$
$$\cdot\left[1-G\left(\overline{es}_{p,m+1}\right)\cdot H_{+2}\left(R_{+1},\varnothing\right)\right]\cdot\left[G\left(\overline{es}_{p,m+2}\right)\cdot H_{+1}\left(\varnothing\right)\right].$$

The full process may be demonstrated in the relatively straightforward calculation of Condition 6: (+4) probability of reaction. Noting that $\overline{es}_H$ occurs with probability of 1 in Condition 6, and that

$$\overline{es}_{H,p=2,m=0} = 0.780\cdot ln\left(e\cdot\frac{0+1.147}{1.147}\right) = 0.780$$

$$\overline{es}_{H,p=2,m=1} = 0.780 \cdot ln\left(e \cdot \frac{1+1.147}{1.147} \right) = 1.269$$

$$\overline{es}_{H,p=2,m=2} = 0.780 \cdot ln\left(e \cdot \frac{2+1.147}{1.147} \right) = 1.567.$$

Using Table 3.8 with $Q(2|3|1) = \frac{10}{17}$, resulting in $p = 2$, $m = 1$, and noting for Condition 6 that $P(NR_{+2}, NR_{+1}) = 1$,

$$P_{Q(2|3|1)}\left(R_{+4} \mid NR_{+3}, NR_{+2}, NR_{+1} \right)$$

$$= G\left(\overline{es}_{p=2,m=1} \right) \cdot H_{+4}\left(\{NR_{+3}, NR_{+2}, NR_{+1}\} \right)$$

$$\cdot \left[1 - G\left(\overline{es}_{p=2,m=2} \right) \cdot H_{+3}\left(\{NR_{+2}, NR_{+1}\} \right) \right]$$

$$\cdot P\left(NR_{+2}, NR_{+1} \right) = \left(1 - e^{-1.269} \right) \cdot e^{ -\left[1.915 \cdot ln\left(\frac{3+1.553}{1.553} \right) \right] }$$

$$\cdot \left[1 - \left(1 - e^{-1.567} \right) \cdot e^{ -1.915 \cdot ln\left(\frac{2+1.553}{1.553} \right) } \right] = 0.076.$$

$$P_{Q(2|3|1)}\left(R_{+4} \mid R_{+3}, NR_{+2}, NR_{+1} \right)$$

$$= G\left(\overline{es}_{p=2,m=1} \right) \cdot H_{+4}\left(\{R_{+3}, NR_{+2}, NR_{+1}\} \right)$$

$$\cdot G\left(\overline{es}_{p=2,m=2} \right) \cdot H_{+3}\left(\{NR_{+2}, NR_{+1}\} \right) \cdot P\left(NR_{+2}, NR_{+1} \right)$$

$$= \left(1 - e^{-1.269} \right) \cdot e^{ -\left[1.298 \cdot ln\left(\frac{3+1.553}{2+1.553} \right) + 1.915 \cdot ln\left(\frac{2+1.553}{1.553} \right) \right] }$$

$$\cdot \left(1 - e^{-1.567} \right) \cdot e^{ -1.915 \cdot ln\left(\frac{2+1.553}{1.553} \right) } = 0.017$$

Similarly, from Table 3.8, we have $Q(2|3|0) = \dfrac{7}{17}$, so

$$P_{Q(2|3|0)}\left(R_{+4} \mid NR_{+3}, NR_{+2}, NR_{+1}\right)$$

$$= G\left(\overline{es}_{p=2,m=0}\right) \cdot H_{+4}\left(\{NR_{+3}, NR_{+2}, NR_{+1}\}\right)$$

$$\cdot \left[1 - G\left(\overline{es}_{p=2,m=1}\right) \cdot H_{+3}\left(\{NR_{+2}, NR_{+1}\}\right)\right]$$

$$\cdot P\left(NR_{+2}, NR_{+1}\right) = \left(1 - e^{-0.780}\right) \cdot e^{-\left[1.915 \cdot ln\left(\frac{3+1.553}{1.553}\right)\right]}$$

$$\cdot \left[1 - \left(1 - e^{-1.269}\right) \cdot e^{-1.915 \cdot ln\left(\frac{2+1.553}{1.553}\right)}\right] = 0.059.$$

$$P_{Q(2|3|0)}\left(R_{+4} \mid R_{+3}, NR_{+2}, NR_{+1}\right)$$

$$= G\left(\overline{es}_{p=2,m=0}\right) \cdot H_{+4}\left(\{R_{+3}, NR_{+2}, NR_{+1}\}\right) \cdot G\left(\overline{es}_{p=2,m=1}\right)$$

$$\cdot H_{+3}\left(\{NR_{+2}, NR_{+1}\}\right) \cdot P\left(NR_{+2}, NR_{+1}\right) = \left(1 - e^{-0.780}\right)$$

$$\cdot e^{-\left[1.298 \cdot ln\left(\frac{3+1.553}{2+1.553}\right) + 1.915 \cdot ln\left(\frac{2+1.553}{1.553}\right)\right]} \cdot \left(1 - e^{-1.269}\right)$$

$$\cdot e^{-1.915 \cdot ln\left(\frac{2+1.553}{1.553}\right)}$$

$$= 0.012.$$

$$Condition\ 6: (+4)\ P\left(R_{+4}\right) = \frac{10}{17} \cdot P_{Q(2|3|1)}\left(R_{+4}\right) + \frac{7}{17} \cdot P_{Q(2|3|0)}\left(R_{+4}\right)$$

$$= \frac{10}{17} \cdot 0.093 + \frac{7}{17} \cdot 0.071 = 0.084.$$

Besides becoming increasingly cumbersome, this process is likely to degrade in accuracy further from the intrusion point due to factors such as decreased likelihood of observing the intrusion, the observing member of interest missing the reaction of some of the intermediate members, or possibly the experimenter mistaking some reactions which seem possible for Condition 4: (+1) which was busiest from a queue member reaction standpoint. This approach does work quite well for the data evaluated, particularly for empirical reaction probabilities greater than 0.1, but further validation is necessary with possible improvements as additional data sets become available. Using this and previous mathematical modeling tools already presented, we can now solve for $\widehat{N}_0$ and $\overline{es}(\widehat{N}_0)$ using already derived Condition 1: (+1), 2: (+2), 5: (+2), and 6: (+3) sensation magnitude results and $H(UI)$ from Equation 3.5.

3.3.5 Estimating $\overline{es}_{H,1,0}$ and $\overline{es}_{H,2,0}$

Using Condition 1: (+1) and Condition 2: (+2) for $\overline{es}_{H,1,0}$ values and Condition 5: (+2) and Condition 6: (+3) for $\overline{es}_{H,2,0}$ values based on $H(UI)$, we have what is needed to derive values for $\widehat{N}_0$ and $\overline{es}(\widehat{N}_0)$ using Theorems 2.4 and 2.5. Conditions 1 and 2 involve a one-unit stimulus, and Conditions 5 and 6 involve two-unit stimuli. They are combined respectively to form the set $P = \{1,2\}$. In other words,

$$\overline{es}_{H,1} = \frac{22 \cdot 0.455 + 24 \cdot 0.455}{46} = 0.455 \text{ and } \overline{es}_{H,2}$$

$$= \frac{20 \cdot 0.78 + 20 \cdot 0.78}{40} = 0.78.$$

Since the cumulative sample size of Conditions 1 and 2 is comparable with Conditions 5 and 6 (i.e., 46 versus 40 respectively), we can better justify applying Theorem 2.5 to find $\widehat{N}_0$ such that

$$\left[0.455 - ln\left(\frac{1+\widehat{N}_0}{\widehat{N}_0}\right) \cdot \frac{1}{1} \cdot \frac{0.780}{ln\left(\frac{2+\widehat{N}_0}{\widehat{N}_0}\right)} \right]$$

$$+ \left[0.780 - ln\left(\frac{2+\widehat{N}_0}{\widehat{N}_0}\right) \cdot \frac{1}{1} \cdot \frac{0.455}{ln\left(\frac{1+\widehat{N}_0}{\widehat{N}_0}\right)} \right] = 0$$

TABLE 3.10 Summary of Empirical Versus Theoretical Results for Schmitt et al. (1992) and Milgram et al. (1986) Field Experiments with Permission from the American Psychological Association

Condition	Position in Queue	# Obs.	Field Experiment Reaction Probability	95% Credibility Interval	Theoretical Reaction Probability
Illegitimate High	(+1)	30	0.600	(0.422, 0.755)	0.599
Illegitimate Low	(+1)	30	0.366	(0.054, 0.298)	0.366
Condition 1	(+1)	22	0.364	(0.197, 0.572)	0.364
1	(+2)	14	0.143	(0.043, 0.405)	0.170
1	(+3)	9	0.0	(0.000, 0.259)	0.100
Condition 2	(+2)	24	0.167	(0.068, 0.361)	0.167
2	(+3)	15	0.0	(0.000, 0.171)	0.104
Condition 3	(+3)	20	0.0	(0.000, 0.133)	0.063
Condition 4	(+1)	23	0.870	(0.676, 0.952)	0.817
4	(+2)	23	0.435	(0.255, 0.633)	0.401
4	(+3)	22	0.091	(0.028, 0.280)	0.252
Condition 5	(+2)	20	0.200	(0.082, 0.419)	0.200
5	(+3)	15	0.0	(0.000, 0.171)	0.094
Condition 6	(+3)	20	0.150	(0.054, 0.363)	0.150
6	(+4)	17	0.118	(0.036, 0.347)	0.084

Solving this for social noise in a New York City queue results in $\widehat{N}_0 = 2.056$. The value for $\overline{es}(2.056)$ is then,

$$\overline{es}(2.056) = \frac{1}{2} \cdot \left[\frac{0.455}{ln\left(\dfrac{1+2.056}{2.056}\right)} + \frac{0.780}{ln\left(\dfrac{2+2.056}{2.056}\right)} \right] = 1.148.$$

TABLE 3.11 Summary Results of Unit Sensation Magnitude Values by Situation

Social Situation	$\overline{es}$
Illegitimate Intrusion – Low-Unit Stimulus Sensation Magnitude	$\overline{es}_L = 0.2869$
Crowd-Gathering (CG) Experiment – New York City Sidewalk	$\overline{es}_{CG} = 0.5890$
Illegitimate Intrusion – Grand Central Rail Ticket Counter	$\overline{es}_H = 1.1477$

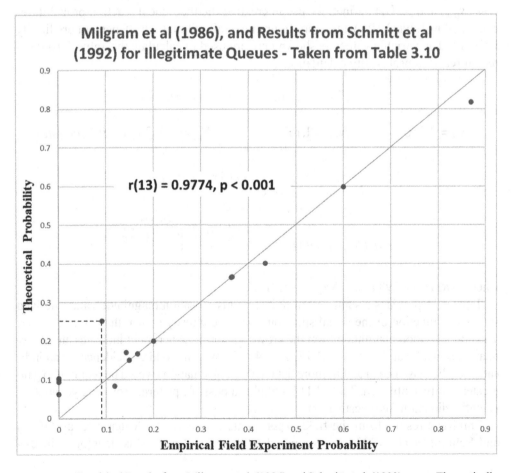

FIGURE 3.1 Empirical Results from Milgram et al. (1986) and Schmitt et al. (1992) versus Theoretically Derived Results for the Same.

Therefore, it should be no surprise using $\widehat{N}_0 = 2.056$ that

$$\overline{es}_{H,1} = 1.148 \cdot ln\left(\frac{1+\widehat{N}_0}{\widehat{N}_0}\right) = 0.455 \ and$$

$$\overline{es}_{H,2} = 1.148 \cdot ln\left(\frac{2+\widehat{N}_0}{\widehat{N}_0}\right) = 0.780, \ for \ \widehat{N}_0 = 2.056.$$

In probability space, when applying these results, $\widehat{N}_0 = 2.056$, and the derived values for $H(UI)$ to Conditions 1: (+1), 2: (+2), 5: (+2), and 6: (+3) a mean square error of 0.0000018 is obtained. This is anticipated but stated to emphasize when viewing Table 3.10 that social queue model parameters were trained on these four conditions and positions. Additional conditions and positions from Milgram et al. (1986) having field experiment reaction probabilities greater than zero could have been used, but these four met Corollary 2.3 requirements and have used the fewest assumptions in their theoretical calculation. Hence, it is felt that their use represents the best that can be obtained with the empirical data available.

The values have been established for $\overline{es}_H, \overline{es}_L. \widehat{N}_0, \widehat{N}_{UA}, \widehat{N}_{UI}$, with the functions $G_{ES}(es_{p,m})$ and $H(UI)$ defined for use in creating the theoretical reaction probabilities contained in Table 3.10. All trial configurations fall within their 95 percent credibility intervals. The variable values as shown here are used to calculate the Table 3.10 theoretical reaction probability calculations.

$$\overline{es}_H = 1.148 \qquad\qquad \overline{es}_L = 0.287;$$

$$\widehat{N}_0 = 2.056, \qquad \widehat{N}_{UA} = 1.147 \qquad\qquad \widehat{N}_{UI} = 1.553 \ (possibly \ constant);$$

$$G_{ES}\left(es_{p,m}\right) = 1 - e^{-es_p \cdot ln\left(e \cdot \frac{m+N_{UA}}{N_{UA}}\right)}$$

$$and \ H_{|UI|+1}(UI) = e^{-\sum_{i=0}^{|UI|-1} k(UI_{i+1}) \cdot ln\left(\frac{|UI|-i+noise}{|UI|-(i+1)+noise}\right)}$$

where $k(R_{i+1}) = 1.298$ and $k(NR_{i+1}) = 1.915$.

Table 3.11 provides a summary of unit stimulus sensation magnitude values derived so far as a function of the social situation. We have already shown that $\overline{es}_H \cong 4 \cdot \overline{es}_L$. The unit sensation magnitude from the crowd-gathering experiment indicates an integer relation as well, namely $\overline{es}_H \cong 2 \cdot \overline{es}_{CG} \cong 4 \cdot \overline{es}_L$. Whether odd coincidence, or an indication of the existence of JNDs, more data in varying situations would be required to gain greater clarity. Until then, Table 3.11 highlights a possible pattern, providing yet another potential direction for investigation.

Empirical results from the field experiments conducted by Milgram et al. (1986) and Schmitt et al. (1992) are compared to theoretical results as developed in this chapter. Figure 3.1 provides a visual depiction of how the two sets of results compare, keeping in mind that a good portion of the empirical reaction probabilities greater than 0.1 were used in determining model variable values. This indicates nothing more than a very good start, with necessary improvements anticipated. Finally, as a marker supporting future effort and comparison, a linear regression performed on the two sets indicates a very strong direct relationship between empirical and theoretical results overall. Note the Pearson's $r(13)$ value of 0.9774 with $p < 0.001$.

With the tools in place and the very strong correlation of calculated results versus empirical results from the two independent queue intrusion experiments, we are now ready to venture outside the queue to determine if the model as derived has any operational utility for social systems that may be transformed into a queue-like hierarchical process. Before doing so, it would be useful to summarize the findings and provide some thoughts on the results before proceeding.

3.4 Summary

Mathematical tools, developed in Chapters 1, 2, and 3, are to be carried forward for evaluation against more socially diverse situations that go beyond the queue as a basic social system. As indicated at the beginning of this book, modeling the queue as a social system provides a means to understand a particular social subgroup based on its social norms. With this understanding, we may be able to better interpret more complex social interactions within the culture of interest, in the present case the western culture with its own social norms, and build on what is developed here. Chapter 4 will begin to consider social scenarios outside of the queue, scenarios which may be transformed into a queue-like process for analysis.

3.4.1 Observations

There are three observations to be made before continuing. The first is the potential progression of development and structure of the derived model as presented. The second involves additional thoughts regarding the Golden Ratio and its appearance in the Milgram et al. (1986) data. The third and final observation is how modeling of the social queue may also apply to more complex social situations, such as may be found in politics and political identification. These are all observations highlighted for future consideration as more is learned about human brain development and operation, and more experimental data having an underlying queue structure becomes available.

As the first observation, the model as derived may represent three distinct stages of complexity. The first, under uniform encoding, accommodates a finite number of social stimuli and accounts for group pressure to support enforcement of social norms. Uniform encoding may have been all that was necessary in prehistoric times with relatively small bands of hunter-gatherers (Dunbar and Sosis, 2018). As human populations grew and became concentrated in city states, the transition from uniform encoding to exponential encoding may have become necessary as a result of social evolution. The transition allows for the existence of infinitely many stimuli and just as many group members to amplify the sensation magnitude caused by the stimuli. The third and final stage is represented by the development of an independent means to gain uncertainty-based information through member reactions that reduce uncertainty. When combined, it seems human evolution may have shifted or may be shifting focus from individual organism survival to group organism survival enhancements. In other words, the group functioning as a larger organism has a better chance of survival than that of the individual. Maybe this is reading too much into what has been observed during model development, but it does seem to warrant further consideration, given the work on multilevel selection theory by Wilson (2015, pp. 47–48) and Cheng (2020).

Equation 3.5 brings us back to Stanley Stevens' power law and the second observation. What seems logical in exponential space may seem more interesting in power space. In this case, recalling the Golden Ratio equals 1.618, and that its inverse is 0.618, consider that

$$\left(\frac{1.618}{1+1.618}\right)^4 = (0.618)^4 = 0.146 \cong e^{-z(3)} = 0.127;$$

$$\left(\frac{1.618}{1+1.618}\right)^3 = (0.618)^3 = 0.236 \cong e^{-z(2)} = 0.205;$$

$$\left(\frac{1.618}{1+1.618}\right)^2 = (0.618)^2 = 0.382 \cong e^{-z(1)} = 0.386$$

These observations remind us of Benjafield and Adams-Weber (1976) and Gross and Miller (1997) and their hypothesis that the Golden Section plays a role in human processing of information. Applying this to Arons and Irwin (1932), in their experiment, they had subjects indicate if a second subsequent weight was lighter, the same, or heavier than a standard weight provided for comparison. In all cases, without the subjects knowing, the second weight was the same as the standard weight, based on 2,500 trials, subject H indicated with a 0.2628 probability the second weight was lighter, with a 0.3156 probability the second weight weighed the same, and with a 0.4216 probability the second weight was heavier. Interestingly enough

$$\frac{0.4216}{0.2628} = 1.604 \cong 1.618.$$

For the third observation, if the postulate is correct that observation of group member reactions is performed by the member of interest to reduce uncertainty regarding the severity of the event, and given that most queues in Milgram et al. (1986) having a theoretical probability of reaction of about 0.1 or less had no reactions directed at the intruder during the experiment, then if four group members in sequence immediately behind the intrusion point do not react, no one remaining behind those four are likely to react. This is shown by,

$$\left(\frac{1.618}{1+1.618}\right)^2 \cdot \left(\frac{1+1.618}{2+1.618}\right)^2 \cdot \left(\frac{2+1.618}{3+1.618}\right)^2 \cdot \left(\frac{3+1.618}{4+1.618}\right)^2$$

$$= \left(\frac{1.618}{4+1.618}\right)^2 \cong e^{-z(4)} = 0.0874$$

This should certainly be true after the fifth member resulting in $e^{-z(5)} = 0.064$. This supports the localized nature of reactions as commented on in Milgram et al. (1986, pp. 688–689) and indicates that multiple members become involved, implying that it is not up to just one or two members to inform the rest of the group. What this also means is that if three of four levels of hierarchy responsible for informing the remainder of the group do not react intentionally for whatever purpose, they can for the most part control uncertainty-based information used by the other in-group members they are responsible to. Control of information, by reaction, non-reaction, or other means, relates to control of perception and how a cohesive group interprets an event. If the larger group is not cohesive, then consider the more cohesive and smaller subgroups whose members will depend on the social attractiveness of the leaders they choose within a hierarchical structure for interpretation of the event.

If social leaders can form and control a highly attractive (i.e., cohesive) group of members who share similar views, then the group and its beliefs can become self-perpetuating as alluded to by Festinger (1957, pp. 186–187, with permission from Stanford University Press),

> *What is more, the degree to which the persistent deviant was rejected was greater in the high attraction than in the low attraction groups. In other words, there was evidence that reduction of dissonance was attempted through rejection of the person who voiced disagreement and that the extent to which this occurred depended upon the magnitude of the dissonance created by this disagreement.*

This chapter finishes the queue model derivation. To investigate possible applications of this model in more complex social situations, we now move outside of the queue and consider social systems that may be transformed back into a queue-like social structure, some of which have already been alluded to. To be a little mischievous, this transformation approach will be termed the queue transform, given it transforms, in appropriate cases, one social system into another. This is similar in nature to the Constant-Q transform used in signal processing used to transform data series information into the frequency domain where the information is more meaningfully and easily evaluated.

References

Arons, L., & Irwin, F. (1932). Equal Weights and Psychophysical Judgements. *Journal of Experimental Psychology*, *15*(6), 733–751. https://doi.org/10.1037/h0070521

Beasley, R. (2016). Dissonance and Decision-Making Mistakes in the Age of Risk. *Journal of European Public Policy*, *23*(5), 771–787. https://doi.org/10.1080/13501763.2015.1127276

Benjafield, J., & Adams-Weber, J. (1976). The Golden Section Hypothesis. *British Journal of Psychology*, *67*(1), 11–15. https://doi.org/10.1111/j.2044-8295.1976.tb01492.x

Burnes, B., & Cooke, B. (2013). Kurt Lewin's Field Theory: A Review and Re-Evaluation. *International Journal of Management Reviews*, *15*, 408–425. https://doi.org/10.1111/j.1468-2370.2012.00348.x

Cheng, J. (2020). Dominance, Prestige, and the Role of Leveling in Human Social Hierarchy and Equality. *Current Opinion in Psychology*, *33*, 238–244. https://doi.org/10.1016/j.copsyc.2019.10.004

Clark, R., & Word, L. (1972). Why Don't Bystanders Help? Because of Ambiguity? *Journal of Personality and Social Psychology*, *24*(3), 392–400. https://doi.org/10.1037/h0033717

Dehaene, S. (2003). The Neural basis of the Weber-Fechner Law: A Logarithmic Mental Number Line. *TRENDS in Cognitive Sciences*, *7*(4), 145–147. https://doi.org/10.1016/S1364-6613(03)00055-X

Dehaene, S., Izard, V., Spelke, E., & Pica, P. (2008). Log or Linear? Distinct Intuitions of the Number Scale in Western and Amazonian Cultures. *Science*, *320*, 1217–1220. https://doi.org/10.1126/science.1156540

Dunbar, R., & Sosis, R. (2018). Optimizing Human Community Sizes. *Evolution and Human Behavior*, *39*(1), 106–111. https://doi.org/10.1016/j.evolhumbehav.2017.11.001

Efthimiou, C. (2010). *Introduction to Functional Equations*. University of Central Florida. Retrieved from www.msri.org/people/staff/levy/files/MCL/Efthimiou/100914book.pdf

FeldhamHall, O., & Shenhav, A. (2019). Resolving Uncertainty in a Social World. *Nature Human Behavior*, *3*, 426–435. https://doi.org/10.1038/s41562-019-0590-x

Festinger, L. (1957). *A Theory of Cognitive Dissonance*. Stanford University Press. Retrieved from www.sup.org/books/title/?id=3850

Festinger, L. (1962). Cognitive Dissonance. *Scientific American*, *207*(4), 93–106. https://doi.org/10.1038/scientificamerican1062-93

Gross, S., & Miller, N. (1997). The "Golden Section" and Bias in Perceptions of Social Consensus. *Personality and Social Psychology Review*, *1*(3), 241–271. https://doi.org/10.1207/s15327957pspr0103_4

Harmon-Jones, E., Harmon-Jones, C., & Levy, N. (2015). An Action-Based Model of Cognitive Dissonance Processes. *Current Directions in Psychological Science*, *24*(3), 184–189. https://doi.org/10.1177/0963721414566449

Jarcho, J.M., Berkman, E.T., & Lieberman, M.D. (2011). The Neural Basis of Rationalization: Cognitive Dissonance Reduction during Decision-Making. *Social Cognitive and Affective Neuroscience*, *6*, 460–467. https://doi.org/10.1093/scan/nsq054

Kitayama, S., Chua, H., Tompson, S., & Han, S. (2013). Neural Mechanisms of Dissonance: An fMRI Investigation of Choice Justification. *NeuroImage*, *69*, 206–212. https://doi.org/10.1016/j.neuroimage.2012.11.034

Klir, G. (2006). *Uncertainty and Information: Foundations of Generalized Information Theory*. John Wiley & Sons, Inc. https://doi.org/10.1002/0471755575

Lewin, K. (1997). *Resolving Social Conflicts and Field Theory in Social Science* (G.W. Lewin, Ed.). American Psychological Association (Kindle Edition). Retrieved from www.apa.org/pubs/books/4318600

Maikovich, A. (2005). A New Understanding of Terrorism Using Cognitive Dissonance Principles. *Journal for the Theory of Social Behavior*, *35*(4), 373–397. https://doi.org/10.1111/j.1468-5914.2005.00282.x

Mann, L. (1969). The Waiting Line as a Social System. *American Journal of Sociology*, *75*(3), 340–354. https://doi.org/10.1086/224787

Melamed, D., Savage, S., & Munn, C. (2019). Uncertainty and Social Influence. *Sociological Research for a Dynamical World*, *5*, 1–9. https://doi.org/10.1177/2378023119866971

Milgram, S., Liberty, J., Toledo, R., & Wackenhut, J. (1986). Response to Intrusion into Waiting Lines. *Journal of Personality and Social Psychology*, *51*(4), 683–689. https://doi.org/10.1037/0022-3514.51.4.683

Miller, R.S. (2001). On the Primacy of Embarrassment in Social Life. *Psychological Inquiry*, *12*(1), 30–33.

Parsons, T. (1951). *The Social System*. Free Press.

Pettigrew, T.F. (1986). The Intergroup Contact Hypothesis Reconsidered. In M. Hewstone & R. Brown (Eds.), *Contact & Conflict in Intergroup Encounters* (pp. 169–195). Basil Blackwell (John Wiley & Sons).

Randles, D., Inzlicht, M., Proulx, T., Tullett, A., & Heine, S. (2015). Is Dissonance Reduction a Special Case of Fluid Compensation? Evidence That Dissonant Cognitions Cause Compensatory Affirmation and Abstraction. *Journal of Personality and Social Psychology*, *108*(5), 697–710. https://doi.org/10.1037/a0038933

Schmitt, B., Dube, L., & Leclerc, F. (1992). Intrusions Into Waiting Lines: Does the Queue Constitute a Social System? *Journal of Personality and Social Psychology*, *63*(5), 806–815. https://doi.org/10.1037/0022-3514.63.5.806

Smith, D., Eggen, M., & St. Andre, R. (2015). *A Transition to Advanced Mathematics* (8th ed.). Cengage Learning.

Stone, J. (2015). *Information Theory: A Tutorial Introduction*. Sebtel Press (Kindle Edition). https://doi.org/10.48550/arXiv.1802.05968

Tanner, T., Haller, R., & Atkinson, R. (1967). Signal Recognition as Influenced By Presentation Schedules. *Perception and Psychophysics*, *2*(8), 349–358. https://doi.org/10.3758/BF03210070

Wilson, D. (2015). *Does Altruism Exist?* Yale University Press.

Winston, W. (1987). *Operations Research: Applications and Algorithms*. Duxbury Press.

4

APPLYING THE QUEUE TRANSFORM

[O]ur social actions are steered by the position in which we perceive ourselves and others within the total social setting.

Lewin, (1945/1997, p. 51) with permission
from the American Psychological Association

The theoretical basis for quantitatively modeling the western first come, first serve queue was derived in Chapters 1–3. The objective in this chapter is to consider the applicability of the derived model against slightly more complex basic social systems with underlying queue-like structures. In mathematics, this would be comparable to the Erlang density function with its underlying Markovian structure. Since the queue is just one of many subgroups in social space, it is not intended nor is it likely to represent all social systems – only a very few simple ones. What is of interest is what it can represent. This chapter evaluates various social psychology experiments utilizing social systems with underlying queue-like structures. Those that have such a structure will be defined as being queue transformable.

Field and laboratory experiments are used in this chapter for their controlled situations and documented results by professional social psychologists. Some of the experiments are more controversial than others, but they all offer insight into efficacy of the queue transform. The queue transform is applied to these experiments to find which are explainable under the transform and which are not. The intended result is a better understanding of how and when the queue transform may be applied. The format used in the analysis of each experiment is background and scenario, original conclusions, application of the queue transform, and discussion of queue transform results. Experiments evaluated under the queue transform along with their authors are:

1. Variations of the Obedience Study – experiments by Milgram (1963, 1965a, 1965b)
2. Lady and a Flat Tire – a field experiment by Bryon and Test (1967)

DOI: 10.4324/9781003325161-5

3. Smoke from a Vent – an experiment by Latane and Darley (1968).
4. Lady Needing Help – an experiment by Latane and Rodin (1969)
5. Accident Victim – an experiment by Clark and Word (1972)
6. Theft of Beer and Money – Experimental design lessons learned from two experiments by Latane and Elman (Latane and Nida, 1981) and Latane and Darley (1969)

To derive theoretical results for the non-queue social systems with queue-like structure, using what has been developed up to this point as a foundation, a proposition must be introduced. In preparation, define the group member social space K such that $P(K) = 1$. Maintaining the nomenclature of Klir (2006, p. 7), let $U_{Pr} \subseteq K$ represent the uncertainty caused by a socially deviant stimulus which a subject of interest might react to for the given social situation. Then $P(U_{Pr}) = G_{ES}(es_{p,m})$ represents the a-priori probability of reaction based on the level of uncertainty felt by the subject of interest. As in Chapter 3, let UI represent the set of uncertainty-based information provided by group members as observed by the subject of interest who is also a member. Now let $U_{Po} \subseteq K$ represent the posterior uncertainty felt by the subject of interest after observing and incorporating uncertainty-based information from other members of his or her group. Define this posterior probability of reaction $P(U_{Po}) = G_{ES}(\overline{es}_{p,m}) H_{|UI|+1}(UI)$. As per basic set theory, assuming dissonance-reducing information $UI = \{\varnothing, NR_1, NR_2, ..., NR_{|UI|}\}$, it was shown in Chapter 3 that for subject member of interest probability of reaction $P(U_{Po})$,

$$U_{Po} = U_{Pr} - \bigcup_{j=1}^{|UI|} UI_j = U_{Pr} \cap \left[\bigcup_{j=1}^{|UI|} UI_j \right]^c = U_{Pr} \cap \left[\bigcap_{j=1}^{|UI|} UI_j^c \right] \text{ and}$$

$$P\left(U_{Pr} \cap \left[\bigcap_{j=1}^{|UI|} UI_j^c \right] \right) = G_{ES}(\overline{es}_{p,m}) \cdot P\left(\bigcap_{j=1}^{|UI|} UI_j^c \right)$$

$$= \left(1 - e^{-\overline{es}_{p,m}} \right) \cdot H_{|UI|+1}(UI)$$

This brings us right back to a slightly more general form of Equation 3.2. To reiterate from Chapter 3, $P\left(\bigcap_{j=1}^{|UI|} UI_j^c \right) = H_{|UI|+1}(UI)$ is the proportion of uncertainty-based dissonance-reducing information available to remove prior uncertainty caused by the observed stimulus event for the given social situation. An example case is depicted in Figure 4.1, where $U_{Pr} = U_{Po} - \bigcup_{i=1}^{3} NR_i$. The correspondence with Klir (2006, p. 7) has now been sufficiently developed to introduce a general proposition to address more complex social systems having underlying queue-like structures using some of the nomenclature Klir developed (i.e., an effort is being taken throughout to recognize related efforts for further investigation as desired).

Proposition 4.1: Model for Basic Social Systems with Underlying Queue-like Structures

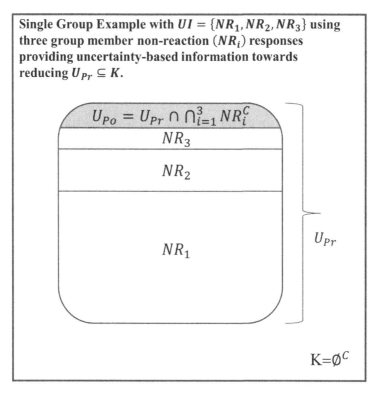

Single Group Example with $UI = \{NR_1, NR_2, NR_3\}$ using three group member non-reaction (NR_i) responses providing uncertainty-based information towards reducing $U_{Pr} \subseteq K$.

$$U_{Po} = U_{Pr} \cap \cap_{i=1}^{3} NR_i^C$$

NR_3

NR_2

NR_1

U_{Pr}

$K = \emptyset^C$

FIGURE 4.1 Conceptual Depiction of Uncertainty and Uncertainty-based Information in Situational Social Space.

Define the reaction group as the subjects of interest and any confederates that the subjects of interest consider equally likely to react. Assumptions currently made for model use are,

1. The social situation is infrequent (no pattern) and ambiguous.
2. That there are one or more simultaneous independent groups (IG), each with independent UI, which share the same k subject(s) of interest, the same social situation, social norms, social stimulus/stimuli p, and any amplification members m relating to $\overline{es}_{p,m}$.
3. Uncertainty-based information is independent of prior uncertainty.
4. Subjects of interest and pertinent group members must have physical or visual presence with one another to gain verifiable reaction/non-reaction uncertainty-based information.
5. Amplification of sensation magnitude occurs when the subject of interest identifies with and feels responsibility toward group members m who model (e.g., social priming) the expected social behavior and/or would consider it the subject of interest's role to react to the stimuli.
6. All k naive subjects of interest are considered equally capable of reducing or eliminating the social deviation by reaction.
7. The dissonance-causing social situation concludes when at least one of the k naive subjects of interest react to the social stimulus/stimuli.
8. If all k subjects of interest are close friends (i.e., not strangers), then *k approaches* 1.

Given these assumptions, the general equation representing probability of at least one of k subject members of interest reacting $P(R)$ in the group is given by,

$$P(R) = P(U_{Po}) = 1 - \left[1 - P\left(U_{Pr} \cap \left[\bigcup_{IG}\left(\cap_{j=1}^{|UI|} UI_j^c\right)\right]\right)\right]^k$$

$$= 1 - \left[1 - P(U_{Pr}) \cdot P\left(\bigcup_{IG}\left(\cap_{j=1}^{|UI|} UI_j^c\right)\right)\right]^k \qquad \text{Equation 4.1}$$

$$= 1 - \left[1 - G_{ES}\left(\overline{es}_{p,m}\right) \cdot P\left(\bigcup_{IG}\left(\cap_{j=1}^{|UI|} UI_j^c\right)\right)\right]^k$$

Proof: The following is a direct proof which assumes all stated associated assumptions are met. For each of the k subjects of interest, each uses the other for uncertainty information to reduce his or her own dissonance. Given IG, the number of independent groups, with U_{pr} the same for each group under Assumption 2, and assuming each of the k subjects of interest are strangers to one another and hence independent, the probability of one or more of the subjects of interest reacting $P(R)$ is,

$$P(R) = P\left(\bigcup_{i=1}^{k} U_{i,Po}\right) = 1 - \prod_{i=1}^{k}\left[1 - P(U_{i,Po})\right]$$

$$= 1 - \left[1 - G_{ES}\left(\overline{es}_{p,m}\right) \cdot P\left(\bigcup_{IG}\left(\cap_{j=1}^{|UI|} UI_j^c\right)\right)\right]^k \qquad \blacksquare$$

The implications of Proposition 4.1 are significant. In essence, the more independent social groups a subject of interest identifies with through various social means, whose members are experiencing the same socially deviant event, the greater the probability of a reaction to create change leading to the reduction of the emotional discomfort caused by the dissonance-causing event affecting them. How this may relate to change and adaptation in an increasingly chaotic environment (Hubler and Pines, 1994, p. 343) caused by increasing uncertainty through active agents who either provide strategic active accurate information or misinformation is a branch topic that is beyond the scope of this effort but certainly has the potential for further exploration in an open forum.

Finally, for a social group composed of long-term friends, it will be shown in Section 4.4 that it is as if there is only one subject of interest (i.e., this is equivalent to $k = 1$ *in Equation* 4.1). Realistically, this is not a completely practical assumption, given variations in each individual's social background, but for small groups of two or three close friends, the actual differences in their social parameter sets directing perception, interpretation, and reaction to various social stimuli are assumed to overlap sufficiently so as to provide a good approximation.

4.1 Variations of the Obedience Study

Background and Scenario: Milgram (1963, 1964, 1965a, 1965b, Experimental design synopses and results with permission from the Estate of Alexandra Milgram and SAGE Publishing). This is Dr. Stanley Milgram's most famous set of experiments, and quite possibly his most controversial. After the Second World War, the world was appalled by and questioned how anyone could do what the Nazis did from 1933 to 1945, where millions of innocent civilians were slaughtered on command. What was learned from his obedience experiments is unsettling at best. Certain controversial aspects of this set of experiments have been touched upon by Patton (1977) and others, but historical data would tend to support the results that Stanley Milgram reports. The experiment begins with an advertisement for volunteers from around the local region, stating only that Yale University is conducting a scientific study on memory and learning. Necessary qualifications are listed along with the statement that if accepted into the study, volunteers will receive payment in the amount of $4.50. At the end of the advertisement, volunteers provide their contact information under the agreement that they want to participate in the study. Already, at this point in the process, a choice has been made whether to participate or not participate, and as Patton (1977) indicates, a study such as this may be more attractive to those looking for acceptance or, alternatively, desiring to be affiliated with a Yale University study given its associated academic prestige. This is important, since once the choice is made, the potential volunteer will begin to find additional reasons or information to internally enhance the importance of participation in the study in an effort to reduce any concerns about the choice taken, in effect, solidifying the choice made (Festinger, 1957, pp. 32–34) and making it more difficult to reverse later. This may imply that a subgroup of the overall population volunteered, but the world is made up of subgroups, so the insight provided in these experiments is still important.

In the original base-line experiment, after being accepted into the study, a naive subject and trained confederate were paid to perform what they were told was a study regarding the effects of punishment on memory. The subjects were told the money was theirs to keep no matter what happened after they arrived at the Yale University laboratory. Once there, the confederate was always chosen as the learner in what appeared to the naive subject as a random selection process. The subject of interest therefore became the teacher under the direction of the experimenter.

The baseline remote learner experiment (Milgram, 1963) has the learner in an adjacent room from the teacher and experimenter. The remote learner is strapped to a chair with electrodes, then attached, and is instructed to provide the correct response to word pairs provided by the teacher. If the learner responds incorrectly, the teacher provides a shock to the learner via an apparatus in front of him or her consisting of 30 switches, starting at 15 volts, and in 15-volt increments ending at 450 volts. After a wrong answer and shock, the teacher moves to the next word pair and administers the next higher voltage if the learner answers incorrectly. At 300 volts, which ends the group of switches labeled "Intense Shock" on the teacher's machine, the learner pounds on the wall and does not answer any further questions. At this point, the teacher is instructed by the experimenter to consider no response as a wrong answer. At 315 volts, which begins the group of switches labeled "Extremely Intense Shock" on the teacher's machine, the learner again pounds the wall, after which no further pounding is heard. The switches were divided and labeled into a total of seven groups starting with "Slight Shock" and ending with "Danger: Severe Shock." The final two switches after the "Danger: Severe Shock" group were marked "XXX." In fact, there was no shock to the learner, but the teacher (naive subject) from all appearances did not know that.

During the process, if the naive subject in the role of teacher expresses discomfort or unwillingness to continue the experiment, the experimenter, in a calm but firm tone, instructs the teacher to continue in a graduated authoritative response (Milgram, 1963, p. 374). Apparently, the experimenter was sometimes more forceful in prompting the teacher to continue (Ofgang, 2018), possibly taking some liberty in the scripted graduated responses provided, but the structure of the procedure remains the same.

Eight variations (conditions), with 40 trials each, of the obedience experiment were performed by Dr. Milgram with the same basic structure.

1. <u>Baseline</u>: Remote learner as described in Milgram (1963, pp. 375–376)

   a. There were 14 defiant naive subjects out of 40 defying the experimenter and refusing to continue before reaching the maximum of 450 volts, five of which refused to go above 300 volts.
   b. In a prior pilot study, in the absence of any protest from the learner, virtually all naive subjects reached 450 volts (Milgram, 1965b, p. 61).

2. (Remote learner) Voice feedback from learner heard by naive subject and experimenter. Learner begins to grunt at 75 volts, begins to verbally declare pain of shocks at 120 volts, shouts to be let out and continues to declare pain beginning 150 volts leading to agonized screams beginning 270 volts, states refusal to respond further at 300 volts, and remains silent after 315 volts (Milgram, 1965a, Table 4.1 Base-line, 1965b, footnote 6).

   a. There were 14 defiant naive subjects out of 40 defying the experimenter and refusing to continue before reaching the maximum of 450 volts, a total of 7 refused to go above 150 volts, 8 above 195 volts, and 11 above 300 volts.

3. (Remote learner/two obedient confederates) Voice feedback from learner with two confederate teacher assistants backing the experimenter's directions while assisting the naive subject who administers the shocks (Milgram, 1964, 1965a, Table 4.1).

   a. There were 11 defiant naive subjects out of 40 defying the experimenter and refusing to continue before reaching the maximum of 450 volts, 9 of which refused to go above 300 volts.

4. (Remote learner/two disobedient confederates) Voice feedback from learner with two confederates assisting the naive subject. The first confederate assistant protests and refuses to continue the experiment at 150 volts, the second at 210 volts, with both taking a seat elsewhere in the room as the experiment continues (Milgram, 1964, 1965a, Table 4.1).

a. There were 36 defiant naive subjects out of 40 defying the experimenter and refusing to continue before reaching the maximum of 450 volts, labeled "XXX," 13 of which refused to go above 195 volts and 31 refused above 300 volts.

5. (Learner in visual proximity) Learner is 1.5 feet from naive subject with voice feedback. Experimenter is also in the same room (Milgram, 1965b, p. 62).

a. There were 24 defiant naive subjects out of 40 defying the experimenter and refusing to continue before reaching the maximum of 450 volts, labeled "XXX."

6. (Learner in touch proximity) Naive subject holds learner's hand down on shock plate after 150 volts with experimenter in the room (Milgram, 1965b, p. 62)

a. There were 28 defiant naive subjects out of 40 defying the experimenter and refusing to continue before reaching the maximum of 450 volts, labeled "XXX."

7. (Remote learner and remote experimenter) Remote learner provides voice feedback. Experimenter either leaves the laboratory after giving instructions and continues further communication by phone or is never seen and provides instructions by tape recording (Milgram, 1965b, p. 65).

a. Does not meet assumption 4 of Proposition 4.1. Not analyzed.

8. (Remote learner) Voice feedback, experimenter in the room with naive subject in an alternate social situation. The experiment was conducted outside of the Yale campus in a sparsely decorated three-room office suite in a run-down commercial building in the town of Bridgeport under the guise of a private firm conducting research for industry (Milgram, 1965b, p. 70).

a. There were 21 naive subjects out of 40 defying the experimenter and refusing to continue before reaching the maximum of 450 volts, labeled "XXX."

Original Observations and Conclusions: In his original article addressing the baseline experiment, Milgram (1963) relays two findings and 13 features that are thought important in explaining the obedience observed. The first finding is that 26 subjects abandoned a fundamental moral rule not to hurt another person against his or her will. All that was required for this abandonment was instructions from an authority having no special powers to enforce his commands. The second finding was the observed high level of tension experienced by the subjects. In this case it might be argued that choosing between two conflicting social norms led to increasing dissonance, one generated by the apparent need to follow directions from an authority for what seemed a just cause and the other by the apparent harm being caused to the learner. Of the 13 features discussed, the one that is most pertinent to this effort is the ambiguity of the situation as perceived by the teacher for what seemed a legitimate situation and good cause in a closed environment (Milgram, 1963, pp. 377–378).

The variations as discussed in Milgram (1964, 1965a, 1965b) are more interesting and spill over into a two-group dynamic: one group that violates a social norm for the cause of science at a prestigious institution and the other that upholds the social norm of doing no harm to another human. In trying to understand and convey what he was seeing, Stanley Milgram elaborates on the group affiliation aspect in Milgram (1965a, pp. 131–134). Using the queue transform here, many of the same conclusions are reached. Finally, in reflection, Stanley Milgram provides his view at the end of the article summarizing his obedience experiments, and although all is worth quoting, the first two sentences are provided here.

> The results, as seen and felt in the laboratory, are to this author disturbing. They raise the possibility that human nature, or – more specifically – the kind of character produced in American democratic society, cannot be counted on to insulate its citizens from brutality and inhumane treatment at the direction of malevolent authority.
>
> *(Milgram, 1965b, p. 75, with permission from SAGE Journals)*

Application of the Queue Transform: Before proceeding, the source of dissonance must be defined. In all cases, the source of dissonance is the conflict between two group social norms felt by the naive subject. The first group is represented by the experimenter, a group member of a large and prestigious academic institution that is perceived to be conducting an experiment for an important cause. The second is the larger community group of which the naive subject is a member, a group that has the social norm that it is wrong to harm others. It is the naive subject making the choice at each shock level as to whether to administer shock, and it is that choice, based on group identification and resultant uncertainty information, that regulates the dissonance within the naive subject.

Condition 1 analysis: Given the inverse relationship between tolerated behavior and dissonance magnitude (Zipf, 1948, loc. 10890), since it is considered sadistic to hurt people with electric shocks, assume in the absence of the experimenter that the naive subject would stop administering shocks with probability near 1 prior to reaching 450 volts, such that $G_{ES}\left(\overline{es}_{1,0}\right) \cong 1$. Now, with the experimenter as a non-reacting group member between the learner as an abstraction in the other room and the naive subject, the probability of defying the experimenter who is providing uncertainty information $\{NR_{+1}\}$ is,

$$G_{ES}\left(\overline{es}_{1,\{NR+1\},0}\right) = G_{ES}\left(\overline{es}_{1,0}\right) \cdot e^{-1.915 \cdot ln\left(\frac{1+1.553}{1.553}\right)}$$

$$= e^{-1.915 \cdot ln\left(\frac{1+1.553}{1.553}\right)} = 0.386.$$

Using the empirical probability of $\frac{14}{40} = 0.35$ that the naive subject actually defies the experimenter, the 95 percent credibility interval is $(0.222, 0.506)$ which 0.386 is well within. As an observation to keep in mind as we continue, it is possible that the actual probability of a naive subject actually stopping the experiment in the absence of the experimenter before reaching 450 volts is,

$$G_{ES}\left(\overline{es}_{1,0}\right) = \frac{0.35}{0.386} = 0.907$$

Condition 2 analysis: The learner is still an abstraction verbally protesting from the other room while the experimenter is in the room with the naive subject, possibly interpreting the learner's protests for the teacher (Latane and Nida, 1981, p.309). In this situation, nothing is different from the base-line condition from the standpoint of cumulative probability, so it is not surprising that Conditions 1 and 2 results are the same but now with a 95 percent credibility interval of (0.255,0.46) based on 80 trials. Combining trials with Condition 1 would indicate $G_{ES}\left(\overline{es}_{1,0}\right) < 1$, possibly closer to 0.9 as the maximum likelihood estimator would suggest.

Conditions 3 and 4, with two confederate teacher assistants, are much more interesting. The naive subject labeled Teacher 3 in Milgram (1964, p. 138) remains responsible for administering the electric shock. Since the two confederates have no role in stopping the electric shock, they may be viewed as group members only. Stanley Milgram also considered this as likely for both Condition 3 and 4 experiments (Milgram, 1965a, pp. 132–134).

Condition 3 analysis: Since the two confederates back the experimenter, there is no alternative group for the naive subject to transition to, and therefore the empirical result of $\frac{11}{40} = 0.275$ should be within the credibility interval bounds of Condition 1 and 2 results. It may also be argued the two confederates add cognitive elements (Festinger, 1957, pp. 21–28) to the naive subject supporting social norm violation within the experimenter group. It is argued here the two confederates increased the social noise $\widehat{N}_0$ somewhat thus reducing the probability of reaction (defiance) slightly. Whether bias was inserted by the confederates or not, the impact was not statistically significant.

Condition 4 analysis: Given Axioms 1 and 2, the naive subject will select the most appealing group when the choice presents itself, with the unchosen alternative group becoming the out-group (Tajfel and Turner, 2004, p. 284). As the confederates sequentially express their concern about the experiment to the experimenter, then remove themselves from the experiment by taking a seat at the side of the room, the naive subject observes a number of social factors at play. As Milgram (1965a) noted, the first is that there are minimal consequences, if any, for ending participation, and doing so is not unusual since the two confederates did it. It was also observed that the continued presence of the confederates after leaving could cause embarrassment to the naive subject if he or she continues the shocks, thus causing further dissonance and conflicting with Axiom 2. By the two confederates leaving the team, the naive subject must also assume full responsibility for his or her actions. At this point, the naive subject has two groups to choose from – one supporting the social norm of do no harm to others, the other having the social norm of allowing harm to another for the sake of science.

Noting naive subject reaction is localized around the voltage at which some form of protest by the learner occurs (Milgram, 1963, 1965a) and using Condition 2 for a comparison without confederates,

$$P\left(Condition\ 2\ defies \leq 195\ volts\right) = \frac{8}{40\ trials} = 0.2$$

$$= \left[1 - e^{-es_{1,0}}\right] \cdot e^{-1.915 \cdot ln\left(\frac{1+1.553}{1.553}\right)}$$

Based on this, for Condition 2, $es_{1,0} = 0.73$ at 195 volts. For Condition 4, when one confederate refuses to continue the experiment, the theoretical probability of defiance should be,

$$P\left(Condition\ 4\ defies \leq 195\ volts\right) = \left[1-e^{-0.73 \cdot ln\left(e \cdot \frac{1+1.147}{1.147}\right)}\right] \cdot 0.386 = 0.268$$

The empirical value was $\dfrac{13}{40\ trials} = 0.325$ with a 95 percent credibility interval of (0.201,0.481) within which the theoretical prediction falls comfortably, implying the naive subject is still using the experimenter for uncertainty information. Similarly,

$$P\left(Condition\ 2\ defies < 450\ volts\right) = \frac{14}{40\ trials} = 0.35$$

$$= \left[1-e^{-es_{1,0}}\right] \cdot e^{-1.915 \cdot ln\left(\frac{1+1.553}{1.553}\right)}$$

From this, $es_{1,0} = 2.375$ at 435 volts. Then,

$$P\left(Condition\ 4\ defies < 450\ volts\right) = \left[1-e^{-2.375 \cdot ln\left(e \cdot \frac{2+1.147}{1.147}\right)}\right] = 0.992$$

, or

$$P\left(Condition\ 4\ defies < 450\ volts\right) = 1-e^{-2.375} = 0.907.$$

The Condition 4 empirical probability of defiance is $\dfrac{36}{40\ trials} = 0.90$. What this implies is, if only 90 percent (0.907) of the naive subjects even consider stopping,

as seems to be the case in all conditions to this point, or even without considering the confederate members, if all naive subjects consider stopping as an option, the experimenter is no longer considered a group member to the naive subject and is no longer used for uncertainty information. The naive subject has in effect shifted with the two confederates to the community group where the social norm of do no harm is followed.

Conditions 5 and 6 analysis: Both are basically the same, with possibly a little more intense discomfort in Condition 6 since the naive subject is holding the learner's hand down on the shock plate. In both conditions, the learner is visible and in the same room, creating a two-group social situation. The first group is the same as the Condition 2 group. The second group now uses the learner as the source of uncertainty information and the experimenter as the cause of the naive subject's dissonance. It is proposed that the naive subject found the learner's protests in front of the experimenter to be a means of dissonance reduction, much like that of a queue member in front of the subject member of interest reacting to an intrusion. This interpretation would imply that a naive subject's dissonance caused by leaving the experimenter group is greater than remaining and using the learner to protest the experiment (Axioms 1 and 2). We then have,

$$P(R) = 1 - \left[1 - G_{ES}\left(\overline{es}_{1,0}\right) \cdot P\left(\bigcup_{IG \in \{NR_1, R_1\}} \left(\bigcap_{j=1}^{|UI|} UI_1^c \right) \right) \right]$$

$$= G_{ES}\left(\overline{es}_{1,0}\right) \cdot P\left(NR_1^c \cup R_1^c \right)$$

$$= G_{ES}\left(\overline{es}_{1,0}\right) \cdot \left[e^{-1.915 \cdot ln\left(\frac{1+1.553}{1.553}\right)} + e^{-1.298 \cdot ln\left(\frac{1+1.553}{1.553}\right)} \right.$$

$$\left. - e^{-1.915 \cdot ln\left(\frac{1+1.553}{1.553}\right) - 1.298 \cdot ln\left(\frac{1+1.553}{1.553}\right)} \right]$$

$$= 0.907 \cdot \left[0.386 + 0.525 - 0.386 \cdot 0.525 \right]$$

$$= 0.907 \cdot 0.708 = 0.642.$$

The Condition 5 visual proximity result of 0.6 has a 95 percent credibility interval of (0.45, 0.74). The Condition 6 touch proximity result of 0.7 has a 95 percent credibility interval of (0.545, 0.819). The theoretical result of 0.642 falls well within both. It is mentioned for future consideration that the 0.6 value may be lower due to increased social noise/ambiguity, given some abstraction is still possible in the visual proximity situation.

Condition 8 analysis: Recall that the experimenter is no longer associated with a laboratory at Yale University but instead is a private contractor conducting research for industry. Interpretation is not possible here without an alternate condition for the same social situation. As two possibilities, based on the previous experimental conditions evaluated at the Yale laboratory, it would seem the remote learner is either now part of the naive subject's group and is providing uncertainty information, conflicting with assumption 4, or the sensation magnitude has actually decreased for some reason, possibly due to location or other factors. If the first, then

$$P\left(Condition\ 8\ defies < 450\ volts\right) = 0.907 \cdot e^{-1.298 \cdot ln\left(\frac{1+1.553}{1.553}\right)}$$

$$= 0.907 \cdot 0.525 = 0.476$$

For this experiment, the empirical probability was $\dfrac{21}{40\ trials} = 0.525$ with a 95 per-cent credibility interval of $(0.374, 0.671)$ with the theoretical explanation being well within the credibility interval. Otherwise, if the social situation has changed the sensation magnitude somehow, then the new $es_{1,0}$ value would be $-ln(1-0.525) = 0.744$, not the $\overline{es}_{1,0} = 2.375$ found at Yale. Given that there are no further experiments for this social situation from which to better understand the results, it would seem that this will remain an enigma for now under the queue transform and given assumptions.

Discussion of Queue Transform Results: This was a complex set of experiments, but, given today's social environment, well worth considering. It is clear from these experiments and from real-world events that when combining an authority figure from a respected institution, tailored uncertainty information, and the right social situation, there is in fact an impact on behavior. It is clear that Patton (1977) had some poten-tially valid arguments regarding the obedience experiments, but history as data would indicate that Stanley Milgram highlighted an underlying structure of human behavior that needs to be acknowledged and dealt with if social stability is to ever be achieved. Results from Conditions 1 through 6 would seem to indicate either 10 percent of the sample population has some anti-social challenges to address, or 10 percent saw through the cover story of the experiment. Overall, the six experiments successfully evaluated would indicate we are defined by the group(s) we identify with. Further analysis of experiments such as these and others should improve application of this model.

4.2 Lady and a Flat Tire

Background and Scenario: This experiment by Bryon and Test (1967, pp. 401–403, Experimental design synopsis and results with permission from the American Psycho-logical Association) was intended to study the use of social models and their effect on altruistic behavior. As an example, children are more inclined to donate to a charitable organization if they see an adult model do so first. In this case, the model used was a man helping a woman fix her flat tire by the side of a road in a residential area of Los Angeles, California. Two conditions were evaluated, the first condition involving a confederate standing by her 1964 Ford Mustang with a flat rear tire and an inflated tire leaning visibly against her car. The second condition involved the same scenario, but with a model car stationed along the side of the same road 440 yards before the woman needing help with her flat tire. The model car was raised by a jack with a young female confederate watching a male confederate change the tire. The location of both cars was chosen so that there was no opportunity to avoid passing both the model car and the woman needing help. The experiment was performed on two consecutive Saturdays, using the same two periods of time for data collection each day resulting in a total of four events. Each event involved 1,000 vehicles passing by. Effort was made to mix up the locations to exploit changing traffic patterns. Defining a reaction as a vehicle pull-ing over to help the young lady, the results are:

Condition 1: With Model (Day 1: 1345–1515 and Day 2: 1630–1730) – 2,000 opportunities, 58 reactions

Condition 2: Without Model (Day 1: 1600–1700 and Day 2: 1400–1530) – 2,000 opportunities, 35 reactions

Original Conclusions: It was noted that the time of day had little impact on aid provided. The general belief provided was that observing the helping behavior served as a reminder of the social norm that we should help one another.

Application of the Queue Transform: Since the confederate who is ostensibly in need of help is standing by the side of the road with the tire propped against her car, a passerby might think she was waiting for help already on the way. If she did not seem in distress, which she did not appear to be, then ambiguity is introduced as to whether she needs help. With situational ambiguity introduced, Proposition 4.1 may be applied. To apply the queue transform, assume the following:

1. A passerby does not stop to help if a car in front of him or her does, implying all uncertainty-based information is from intermediate non-reacting cars.
2. Having observed the model situation, and that the lady in distress is within sight of the model, the subject of interest feels increased responsibility to react. This results in $m = 1$ for the "with model" condition.
3. Assume (*speed* · 12.5 *seconds*) equals driver scanning distance. California Department of Motor Vehicles states that drivers should scan 10 to 15 seconds of driving distance ahead (California Driver Handbook, 2020, p. 42). Australian guidance is similar in stating that drivers should scan ahead 12 seconds (Scanning for Hazards, 2022).

Using Equation 4.1 with $IG = 1$ and $m = 1$ when the model is within visual sight of the subject of interest leads to,

$$P(R \mid with\ model) = \frac{58}{2000} = 0.029 = \left(1 - e^{-\overline{es}_{1,1}}\right) \cdot e^{-z(|UI|)}, \text{and}$$

$$P(R \mid no\ model) = \frac{35}{2000} = 0.0175 = \left(1 - e^{-\overline{es}_{1,0}}\right) \cdot e^{-z(|UI|)}.$$

This could be solved directly since there are two equations and two unknowns (i.e., $\overline{es}_1$ and $|UI|$), but let us find the value for m in the "with model" for the sake of argument by solving for $|UI|$ in an independent manner.

Since we do not know if females are equally likely as males to help, or what the proportion of males to females is that drive by, the sensation magnitude $\overline{es}_{1,0}$ for this must be an average value of the two. $P(R)$ will be the average probability a subject pulls over for this given social situation. Given there are a total of 2,000 cars passing by over a total of 150 minutes for each condition and that residential speed limits are 25 miles per hour (mph) unless otherwise posted (California Driver Handbook, 2020,

p. 41), then the number of cars in the quarter mile (0.25 *miles*) between the model car and lady in distress is calculated in the following manner,

$$0.25\ miles \cdot \frac{2000\ cars}{2.5\ hrs \cdot 25\ mph} = 0.25\ miles \cdot 32 \frac{cars}{mile} = 8\ cars.$$

Based on the third assumption, the subject scans 12.5 seconds ahead at a speed of 12.22 yards per second or a decision-making scanning distance of 152.7 yards. There will then be an average of $\frac{152.7\ yards}{440\ yards} \cdot 8\ cars = 2.78$ cars between the subject of interest and the lady in distress when she enters the subject of interest's decision-making scanning distance. Using $|UI| = 2.78$ to derive the no-model condition for $\overline{es}_{1,0}$ leads to

$$\frac{35}{2000} = \left(1 - e^{-\overline{es}_{1,0}}\right) \cdot e^{-1.915 \cdot ln\left(\frac{2.78+1.553}{1.553}\right)},\ or$$

$$1 - e^{-\overline{es}_{1,0}} = \frac{35 \cdot e^{1.915 \cdot ln\left(\frac{2.78+1.553}{1.553}\right)}}{2000} = 0.125,\ so\ that\ \overline{es}_{1,0}$$

$$= -ln(1 - 0.125) = 0.134.$$

It is proposed that the model condition (confederate man helping confederate woman) enforces the social responsibility of the subject in much the same manner as queue members behind the member of interest. The subject has just observed the model and is now observing whether the cars in front of him or her pull over to help. If none of the cars ahead pull over to help, and the subject car feels the good Samaritan at the model car is part of his or her group and situation, the probability the subject car pulls over is now,

$$P(R \mid model) = 1 - \left[1 - \left(1 - e^{-\overline{es}_{1,m}}\right) \cdot e^{-1.915 \cdot ln\left(\frac{2.78+1.553}{1.553}\right)}\right]$$

$$= \left[1 - e^{-0.134 \cdot ln\left(e \cdot \frac{m+1.147}{1.147}\right)}\right] \cdot e^{-1.915 \cdot ln\left(\frac{2.78+1.553}{1.553}\right)}$$

$$= 0.029 = \frac{58}{2000}.$$

Solving for *m* results in

$$m = \frac{1.147}{e} \cdot \left(\frac{0.1402}{0.1402 - 0.029} \right)^{\frac{1}{0.134}} - 1.147 = 1.232 \cong 1.$$

Going back now and using $m = 1$ under assumption 5 of Proposition 4.1 results in

$$P(R \mid model) = \left[1 - e^{-0.134 \cdot ln\left(e \cdot \frac{1+1.147}{1.147} \right)} \right] \cdot e^{-1.915 \cdot ln\left(\frac{2.78+1.553}{1.553} \right)}$$

$$= 0.1959 \cdot 0.1402 = 0.027$$

With an empirical result of 0.029, the 95 percent credibility interval for the "with model" condition is (0.023, 0.037), indicating that the theoretical result of 0.027 is well within the accepted interval. Given the large sample size for this experiment, if we use $\widehat{N}_{UA} = 1.0$, we obtain the theoretical result of 0.0282 indicating $\widehat{N}_{UA} = 1.0$ is potentially more accurate than $\widehat{N}_{UA} = 1.147$ for this situation.

Discussion of Queue Transform Results: The basic reaction probability without the model and without observing others is $1 - e^{-0.134} = 0.125$. This seems rather low, but then we do not know what level of social noise is involved. The relevance to the subject of interest in this situation is small, given that the subject does not know the woman, she is very likely not in his or her immediate (personal) social group, there is potential legal or physical danger when stopping to help a stranger, she does not seem in distress, and her situation has no impact on the subject other than some possible amount of guilt for not stopping. Given the experiment was conducted in a residential section of Los Angeles, the resultant desire to avoid interaction may also be higher than if the experiment were conducted in a rural town (Milgram, 1970).

It is interesting to note in this field experiment that there are drivers both ahead and behind the subject, and there is the model. Drivers ahead and within scanning range of the subject are used as sources of uncertainty information, but drivers behind the subject do not influence his or her probability of reaction, indicating the subject of interest, unlike in a queue, does not perceive cars behind him or her as group members possibly due to the anonymity afforded by driving in a car, or the perception is that those on the road do not expect others to help.

The model seems to represent drivers who identify in varying degrees with a group having the social norm of assisting others or at least assisting a lady in distress. These drivers may consider the Samaritan assisting the woman at the model car as a group member, now behind them, who may judge their reaction or, worse, may end up having to react if the subject does not. Not stopping could cause the subject to feel as though he or she is not supporting this group's social norms.

4.3 Smoke From a Vent

Background and Scenario: Latane and Darley (1968, Experimental design synopsis and results with permission from the American Psychological Association). In the original New York Times 1964 version of the Kitty Genovese murder article, 38 people overheard her struggle and eventual murder without responding or calling the police. Although many of the original details surrounding that brutal, early morning murder of 28-year-old Kitty Genovese in New York City were not entirely accurate (McFadden, 2016), the original story created strong interest as to why people may or may not help another in distress. In their pursuit to understand this, Latane and Darley (1968) conducted an experiment where a mixture of naive subjects and nonreactive confederates were placed in a waiting room and directed to fill out a survey. While filling out the survey, non-toxic smoke created by the experimenters was streamed (puffed) into the room through a wall vent. The probability of reaction metric was simple, either the naive subject went to find a staff member within 6 minutes to report smoke entering the room, or the subject remained in the room filling out the survey. The experiment was conducted using three experimental conditions provided in Latane and Darley (1968, pp. 217–218):

Condition 1: (24 trials) Probability of reaction by one naive subject was 0.75 (18/24).

Condition 2: (10 trials) Probability of reaction by one naive subject and two non-reacting confederates in the room was 0.1 (1/10). In this experimental condition, sometimes both confederates were in the room when the naive subject arrived, sometimes they arrived after the subject, and sometimes one confederate arrived before and one after the subject arrived (Latane and Darley, 1970, p. 47).

Condition 3: (8 trials) Probability of reaction by at least one of three naive subjects was 0.38 (3/8).

Smoke coming from the vent is the unexpected deviation from what should occur for the social situation as described. Given the increasing probability of reaction as a function of time, sensation magnitude associated with the event is also likely influenced by its duration. In leading up to their description and discussion of results for this experiment, Latane and Darley (1968, p. 216) hypothesize that after observing an event, but before acting, a bystander will determine if action is required basing part of that decision process on how others around him or her are reacting.

Original Conclusions: After presenting their findings, alternative hypotheses were offered as to why results over the three conditions occurred as they did. The summary conclusion, based on this experiment and previous studies performed, was (Latane and Darley, 1968, p. 221) that a bystander is influenced by the presence of others, either through their influence, or by the bystander passing responsibility to those around him or her.

In what will be seen as important later, it was also concluded (Latane and Darley, 1968, p. 221) that the relationship between bystanders may be important.

These subjective observations do not explain in a quantitative manner why only one-out-of-ten naive subjects responded in the Condition 2 experiment, nor do they explain the relationship between Conditions 1 and 2, and 1 and 3, but it does turn out, as we shall see, that knowing the personal relationship between bystanders is very important. Latane and Darley (1968) do note that the Condition 2 result may possibly be used to derive the probability of at least one of the three naive subjects in Condition 3 reacting. Their result was $1 - (1 - 0.1)^3 = 0.27$, which means little, given only eight trials were performed resulting in a 0.38 probability of reaction and an incredibly large 95 percent credibility interval of (0.14, 0.70).

Application of the Queue Transform: Dehaene et al. (2008, p. 1218) state that Americans, when shown several dots, count the dots linearly up to about ten, but then exhibit a significant logarithmic component of estimation for sets of dots greater than ten. In Chapter 3, it was observed that in a queue, two people between the intruder(s) and member of interest were treated logarithmically by the member of interest. The member of interest may have counted them as two people, but internal processing of the information based on social impact appears to be logarithmic. Since the assumptions for Proposition 4.1 are met, we may apply the queue transform to the experiment of Latane and Darley (1968).

Condition 1, a single naive subject in the room, is the baseline for this set of experiments. After 6 minutes, the theoretical sensation magnitude for a single stimulus (smoke from vent) is,

$$\overline{es}_{1,0} = -ln(1 - 0.75) = 1.3863$$

Condition 2, having two non-reacting confederates in the room as an ad hoc group, is equivalent to two confederates ahead of the member of interest in a queue having the same unit social stimulus magnitude and social noise as Condition 1. The confederates are now sources of uncertainty information, just as intermediate queue members would be. From this, the Condition 1 result may be applied while introducing two non-reacting queue members between the stimulus and the member of interest. Using Equation 4.1 with $IG = 1$ *and* $k = 1$ results in,

$$\left(1 - e^{-\overline{es}_{1,0}}\right) \cdot e^{-z(2)} = 0.75 \cdot e^{-1.915 \cdot ln\left(\frac{2+1.553}{1.553}\right)} = 0.154.$$

Given only ten trials were conducted in the Condition 2 experiment, resulting in a sample mean probability of reaction of 0.10, the theoretical result of 0.154 is well within the 95 percent credibility interval of (0.023, 0.413).

Condition 3 has three naive subjects who react independently to a common stimulus (i.e., $IG = 1$, $k = 3$) who use the other two members for information. Applying Equation 4.1 leads to,

$$P(R) = 1 - \left[1 - \left(1 - e^{\overline{es}_{1,0}}\right) \cdot e^{-z(2)}\right]^k = 1 - (1 - 0.75 \cdot 0.205)^3$$
$$= 0.394.$$

Recalling the Condition 3 empirical result of 0.38, the theoretical Condition 3 result based on the Condition 1 baseline is almost exact using Proposition 4.1. Given a baseline condition as was provided by Condition 1 of this experiment, the queue transform under Proposition 4.1 seems to adequately model and explain the empirical results obtained in Latane and Darley (1968). What remains to be addressed is the observation that probability of naive member reaction increases with time.

Discussion of Queue Transform Results: There is no known raw data supporting the figure in Latane and Darley (1968, p. 218) so data from their probability of reaction over time is visually approximated for the Condition 1 (single naive subject) and provided in Table 4.1.

What is interesting is the plot of column $\overline{es}_{1,0}(t)$ from which Figure 4.2 is created. As can be observed, the slope is basically linearly increasing until the time when no more reactions occur (i.e., 3.5 minutes). In this case, sensation magnitude for Condition 1 is increasing at an average rate of 0.4 per minute. The question to pose now is whether this increase is from clarifying what the social deviation actually is, assuming relevance remains constant, or possibly the social noise is decreasing through the integration of observation over time using some mental process similar to what is done in narrowband acoustic signal processing.

Latane and Darley (1968, p. 219) note that the median latency for the Condition 1 naive subject to notice smoke entering the room was five seconds; yet, it is over 2 minutes before 50 percent of the subjects report the smoke. Social noise has to do with noticing the event and accounting for any social inhibitions that may impact the desire to react. In Condition 1, as the subject is alone, there would be no inhibitions by other social group members as Latane and Darley (1968, pp. 216–219) discuss for Conditions 2 and 3. It is instead likely there is a delay in reporting just to confirm the event is prolonged and that it will be there when the person it is reported to investigates (i.e., subject of interest desiring to avoid embarrassment).

A simple puff of smoke for one or two seconds could be anything, and, if it did not return, then there would be little immediate reason to report it. Typically, in Condition 1, the subjects investigated the smoke by sniffing it, waiving their hands in front of it, feeling its temperature, etc. After a few more moments of hesitation, the subjects who did report the smoke would calmly walk out and alert a staff member (Latane and Darley, 1968, p. 217). This indicates that the subject's evaluation of the event occurred over time, with a growing concern leading to increasing dissonance. Sensation magnitude of the event eventually levels off even as the event continues. Data from the experiment and/or the experience/creativity of the author are insufficient at this point to come up with a mathematically justifiable explanation.

TABLE 4.1 Probability Naive Subject Reports Smoke vs Time, Latane and Darley (1968, p. 218)

Minutes $t =$	Condition 1: Probability of Report $G_{ES}\left(\overline{es}_{1,0}\right)$	$\overline{es}_{1,0}(t) = -ln\left[1 - G_{ES}\left(\overline{es}_{1,0}\right)\right]$
1	0.375 (9/24)	0.462
2	0.542 (13/24)	0.780
2.5	0.625 (15/24)	0.981
3.5	0.750 (18/24)	1.386
6	0.750 (18/24)	1.386

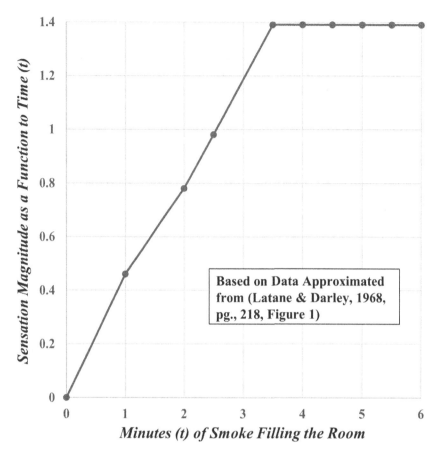

FIGURE 4.2 Sensation Magnitude as a Function of Time from the Introduction of Smoke to Probability of Subject Reacting.

4.4 Lady Needing Help

Background and Scenario: Latane and Rodin (1969, Experimental design synopsis and results with permission from Elsevier Publishing) further explore the effects of social influence by others in the immediate vicinity when someone is observed or heard to need help. In this experiment, the impact on helping is evaluated as a function of whether a pair of students in various conditions will react and help a lady in distress. The importance of this experiment is that it evaluates the reaction of naive subjects who are friends, possibly relating to experiments indicating urban dwellers are less likely to assist than rural dwellers (Milgram, 1970).

The social stimulus in all four conditions is the same. Subjects who volunteered to evaluate games for market research are met by a female representative, who is a confederate in the experiment, and taken to a testing room where they fill out a survey. The representative goes into the next room which is accessible from the testing room. Four minutes after she leaves, and while the student is still filling out the survey, she turns on a high-fidelity tape recorder in the other room. What the subject(s) in the other room hears is a loud crash and scream indicating the representative fell while getting something. They then hear moans indicating she cannot get the shelf off her. After about 130 seconds, if there was no intervention, the representative would limp out and ask the subjects about the noises and if they heard them.

Four conditions are evaluated under this scenario. They are,

Condition 1: (26 trials) One naive subject is greeted and fills out the questionnaire alone in the testing room.

Condition 2: (14 trials) One naive subject and one nonreactive confederate are greeted and taken to the testing room where they fill out the questionnaire.

Condition 3: (20 trials) Two naive student strangers are greeted and taken to the testing room where they fill out the questionnaire.

Condition 4: (20 trials) Two naive student friends are greeted and taken to the testing room where they fill out the questionnaire. Since not mentioned in this experiment, it is worth noting that in the study by Clark and Word (1972), the pair of friends used in that experiment had been friends for an average of 3 years.

Original Conclusions: Results from the experiment by condition and fraction of subjects who react as a function of time are provided in Table 4.2. Probability of reaction fractional values remain constant after 110 seconds until the experiment ends at 130 seconds.

Final conclusions by the study authors concerning this experiment indicate that social inhibition effects may be somewhat general even under various social situations (i.e., smoke-filled room, a lady in distress, etc.). It has been shown now in multiple experiments that bystanders are less likely to react if other bystanders are present, with the nature of the bystander (i.e., confederate or stranger) effecting how likely the probability of reaction is. In this experiment, a nonreactive confederate is the greatest inhibitor, an independent stranger capable of reacting provides a moderate reduction in probability of at least one of the subjects reacting, and a friend the least amount of inhibition when a second subject considers reacting to an ambiguous but potential emergency. Results also point to why larger cities, filled with strangers, may be less safe than smaller towns from the standpoint of getting aid if in distress. In small towns, people tend to know one another, or at least feel a sense of common community. It is then possible that in small towns, even though they may not all be friends, they are not likely to feel like strangers when experiencing an ambiguous and emotionally uncomfortable situation together. Additionally, it is anticipated that group cohesiveness is greater in small towns, amplifying dissonance caused by a social deviation when other town members are in visual proximity.

Application of the Queue Transform: As was done for the smoke-filled room, we must first find the baseline sensation magnitude $\overline{es}_{1,0}$ sample mean. As per Table 4.2, after 70 seconds, the probability of reaction for a single naive subject in this situation stabilizes at $\frac{18}{26} = 0.6923$. Therefore,

$$\overline{es}_{1,0} = -ln(1 - 0.6923) = 1.179.$$

TABLE 4.2 A Lady in Distress Experiment Probability of Reaction Versus Time $P(R, t)$ Data Approximated from Latane and Rodin (1969, pp. 193–195, Figures 4.1 and 4.2)

	10 sec	20 sec	30 sec	40 sec	50 sec	60 sec	70 sec	80 sec	90 sec	100 sec	110 sec
Condition 1: Subject alone $P(R,t)$	$\frac{0}{26}$	$\frac{6}{26}$	$\frac{11}{26}$	$\frac{16}{26}$	$\frac{17}{26}$	$\frac{17}{26}$	$\frac{18}{26}$	$\frac{18}{26}$	$\frac{18}{26}$	$\frac{18}{26}$	$\frac{18}{26}$
Condition 2: Subject w/ confederate $P(R,t)$	$\frac{0}{14}$	$\frac{0}{14}$	$\frac{1}{14}$	$\frac{1}{14}$	$\frac{1}{14}$	$\frac{1}{14}$	$\frac{1}{14}$	$\frac{1}{14}$	$\frac{1}{14}$	$\frac{1}{14}$	$\frac{1}{14}$
Condition 3: Two naive strangers $P(R,t)$	$\frac{0}{20}$	$\frac{0}{20}$	$\frac{1}{20}$	$\frac{1}{20}$	$\frac{3}{20}$	$\frac{5}{20}$	$\frac{5}{20}$	$\frac{6}{20}$	$\frac{7}{20}$	$\frac{7}{20}$	$\frac{8}{20}$
Condition 4: Two friends (naive subjects) $P(R,t)$	$\frac{0}{20}$	$\frac{2}{20}$	$\frac{7}{20}$	$\frac{12}{20}$	$\frac{14}{20}$	$\frac{14}{20}$	$\frac{14}{20}$	$\frac{14}{20}$	$\frac{14}{20}$	$\frac{14}{20}$	$\frac{14}{20}$

As per Proposition 4.1, the probability of the naive subject reacting $P(R)$ in Condition 2 of this study, given that the confederate will not react (i.e., $NR_{Conf} = 1.0$ is

$$P(R) = \left(1 - e^{-\overline{es}_{1,0}}\right) \cdot e^{-z(1)} = 0.692 \cdot 0.386 = 0.267.$$

This seems much larger than the value of 0.071 for a naive subject and confederate combination at 130 seconds. Keeping in mind though that there were only 14 trials conducted in Condition 2, the theoretical result falls well within the 95 percent credibility interval of (0.016,0.319). On the regression plot, the empirical 0.071 result almost appears as an outlier, once again indicating the need to plan for more trials in an effort to reduce credibility intervals to something more meaningful.

Using Proposition 4.1 on Condition 3 of this study, the theoretical probability of at least one of the two naive strangers reacting is

$$P(R) = 1 - \left[1 - \left(1 - e^{\overline{es}_{1,0}}\right) \cdot e^{-z(1)}\right]^2 = 1 - (1 - 0.267)^2 = 0.463.$$

The empirical result for Condition 3 based on 20 trials is 0.40. The associated 95 percent credibility interval is (0.218,0.615) indicating that the theoretical result of 0.463 using the Condition 1 baseline value for $\overline{es}_{1,0} = 1.179$ is well within the credibility interval.

Up to this point, we have found that when one or more confederates or strangers are involved, the queue transform has satisfactorily matched empirical results. It is assumed that this is the case since those two conditions reflect the conditions modeled for the queue as a social system. Having two naive friends in this situation offers a new social dynamic, given the strong personal relationship which implies similarity of perception, interpretation, and reaction (Parkinson et al., 2018). In addition, Latane and Rodin (1969, pp. 197, 200–201) postulate the following regarding the two friends' condition:

1. Unlike with strangers, friends are less likely to fear embarrassment between them.

   a. It was found in Zoccola et al. (2011, p. 927) that the desire to diffuse responsibility to a stranger to avoid embarrassment or responsibility exists and follows the queue transform model. Since mentioning an embarrassing fact, in this case (Zoccola et al., 2011) to the confederate interviewer results in making a choice, it is proposed that dissonance is involved and that Axiom 2 of Festinger (1957) pertains (i.e., avoid increasing the dissonance).

2. Friends are less likely to misinterpret each other's non-reaction than they may with strangers. It was noted by the experimenters that friends seemed better able to convey their concern and plan of action both nonverbally and verbally to one another.

3. Friends are less likely to pass responsibility of action to another friend thereby reducing diffusion of responsibility to the other participant.

4. In Condition 2, 14 percent of the subjects reported a moderate degree of influence by the other's presence, 30 percent reported the same in Condition 3, and 70 percent reported the same in Condition 4 which involved the two naive friends.

To theoretically evaluate Condition 4, where two friends are the naive subjects, we need only apply Proposition 4.1 and note that no uncertainty-based information is gained by observing the friend. Quantitatively stated,

$$P(R) = \left(1 - e^{-\overline{es}_{1,0}}\right) = 1 - e^{-1.179} = 0.692.$$

The empirical probability of reaction for Condition 4 was 0.7, so the nearly identical theoretical result is within the 95 percent credibility interval of $(0.478, 0.854)$ based on 14 reactions out of 20 trials.

Discussion of Queue Transform Results: Use of Proposition 4.1 in this social situation further supports the assumption that friends perceive, interpret, and react to social stimuli similarly for a given social situation as suggested by the work of Parkinson et al. (2018). This might be carried further when comparing rural reactions (closer-knit social groups) to urban reactions (more diverse social groups with varying social norms).

4.5 Accident Victim

Background and Scenario: Clark and Word (1972, Experimental design synopsis and results with permission from the American Psychological Association) build on the work of Latane and Darley by comparing results from situations with no ambiguity to those where ambiguity is present in an experiment designed to focus on the role of social influence among participating group members. It is the second experiment with ambiguity that is of interest here. The procedure in both experiments was that while the student subjects were filling out their questionnaire as directed by the experimenter, a victim posing as a university maintenance employee – who the students had seen in the hallway when reporting to the experiment – would enter the room carrying a ladder and venetian blind. Passing the subjects, he would enter an adjacent room and could be heard working through the closed door. Three minutes after the maintenance man entered the adjacent room, he would push the ladder against the wall and then onto the floor while pulling down the venetian blinds and making it sound as if he fell.

The experiment with ambiguity consisted of two variations, a low ambiguity and high ambiguity event. In the low-ambiguity event, the maintenance man, after the staged accident, would groan and call for help for 75 seconds with decreasing loudness. In the high-ambiguity event, there was no groaning or calling for help, only the crash of the ladder, venetian blinds, and sound of the maintenance man falling to the floor. In both experimental events, students are seated in a room at a table approximately 15 feet from the door which allows access to where the maintenance man was working. Maximum observation time for a reaction by the student is 75 seconds, after which the experiment is terminated. There were three student conditions, all involving naive students. The three conditions follow:

Condition 1: (10 trials) A single naive student is seated at the table filling out the questionnaire. The seat selected is the one closest to the door in which the maintenance man went through.

Condition 2: (10 trials) Two naive students are seated at the table filling out the questionnaire with one student being about two feet further from the door than the seated student closest to the door.

Condition 3: (10 trials) Five naive students are seated at the table filling out the questionnaire. Three of the five student members are even further from the door than the other two who are placed in the same position as Condition 2. Clark and Word (1972, p. 397) indicate the five-person group was actually on average a 4.5-person group.

The response times and probability of reaction, which included either getting up and checking on the maintenance man in the adjacent room or going to inform staff, are contained in Table 4.3.

TABLE 4.3 Mean Reaction Time and Probability of Reaction for the Second Experiment (Clark and Word, 1972, p. 397, With Permission from the American Psychological Association)

Condition Composed of 10 Trials Each	*High-Ambiguity Event Delay, Reaction Probability*
Condition 1. Single Naive Student at the Table	$P(R) = 0.3$
Condition 2. Two Naive Students at the Table	$P(R) = 0.2$
Condition 3. Five Naive Students at the Table	$P(R) = 0.4$

Since there are only ten trials for each event condition, the credibility intervals are large, making meaningful model analysis difficult. The low-ambiguity event is not addressed in this analysis since all probabilities of reaction equal 1. All that may be surmised from the low-ambiguity event is that the sensation magnitude was relatively high, possibly due to lower social noise caused by significantly decreased ambiguity of the situation resulting from the maintenance man groaning. The high-ambiguity event though offers some potentially meaningful insight.

Original Conclusions: Clark and Word (1972) considered what might happen if the modeling approach taken were to consider each student as acting completely independently, as Latane had done in previous experiments. As a result, for the two-student condition

$$1 - (1 - 0.3)^2 = 0.51,$$

and for the five-student condition

$$1 - (1 - 0.3)^5 = 0.832.$$

The two-student result is just within the 95 percent credibility interval of $(0.06, 0.518)$. The five-student result is well outside of its associated 95 percent credibility interval of $(0.167, 0.692)$. Hence, Clark and Word concluded that students were not acting independently, and in both cases the students had an inhibiting influence on one another.

Application of the Queue Transform: Using Proposition 4.1, we have for the single naive student as the baseline case,

$$\overline{es}_{1,0} = -ln(1 - 0.3) = 0.3567.$$

For Condition 2, accounting for each student (stranger) gaining uncertainty-based information from the non-reaction of the other and applying the baseline result to this independent group of two subjects results in at least one student's probability of reacting as

$$1 - \left[1 - \left(1 - e^{\overline{es}_{1,0}}\right) \cdot e^{-z(1)}\right]^2 = 1 - (1 - 0.3 \cdot 0.386)^2 = 0.218.$$

This result is basically the same as the empirical result of 0.2 and is well within the 95 percent credibility interval of $(0.06, 0.518)$. Using this same approach for the five-person (i.e., on average 4.5-person group), it is apparent that something new is happening.

The probability of one student reacting after observing four non-reacting students is

$$\left(1 - e^{\overline{es}_{1,0}}\right) \cdot e^{-z(4)} = 0.3 \cdot e^{-1.915 \cdot ln\left(\frac{4+1.553}{1.553}\right)} = 0.0261.$$

For Condition 3, the probability, under Proposition 4.1, that at least one of the five students reacts is then,

$$1-\left(1-\left(1-e^{\overline{es}_{1,0}}\right)\cdot e^{-z(4)}\right)^{5}=1-\left(1-0.0261\right)^{5}=0.124.$$

As Table 4.3 indicates, the empirical probability of reaction is 0.4. The theoretical probability of reaction of 0.124 is therefore not within the 95 percent credibility interval of (0.167,0.692), indicating it unlikely that Proposition 4.1 used in this manner represents what is occurring socially.

Recall that Clark and Word (1972) state for the five-person group that three of the five were placed even further from the door than the other two. They were concerned about this, certainly from the standpoint of reaction time as they discussed, but possibly for other reasons that were not discussed. It is not clear how much farther away from the two students closest to the door the remaining three were placed, possibly leading to two spatially distinct groups, with one group closer to the door than the other. This might occur where the group of two is on one side of the table, and the second group of three is on the other side. So far, we have shown it is unlikely that the five students were equally influenced by one another as they seem to have been in the two-person group. Instead, it is hypothesized from this analysis that two distinct and independent subgroups formed assuming they were sufficiently separated within the room.

Consider this alternative, where the two students closest to the door adhere to Proposition 4.1 as they did in the two-student condition, but in this case without regard for the other three students further away and seemingly in a physically distinct location from the other two. Under this assumption, the three students furthest from the door act as an independent group without regard for the two students closest to the door. Another way of stating this is that the two students closest to the door form one group, and those furthest from the door form another, with each group being independent and not having the same subjects of interest as required by assumption. Modifying Equation 4.1 to accommodate this leads to

$$1-\left(1-0.3\cdot e^{-1.915\cdot ln\left[\frac{1+1.553}{1.553}\right]}\right)^{2}\cdot\left(1-0.3\cdot e^{-1.915\cdot ln\left[\frac{2+1.553}{1.553}\right]}\right)^{3}$$

$$=1-0.782\cdot 0.827$$

$$=0.354.$$

This alternative is also well within the 95 percent credibility interval. Without knowing the actual placement of students, there may be other variations to consider, but this alternative model provides the closest value to the empirical 0.4 result and therefore is the most likely until more detailed information is obtained, if ever, regarding the experiment's placement of students in the room.

Discussion of the Queue Transform Results: To continue hypothesizing, though fun, would be fruitless given the size of the credibility interval and the lack of detailed experimental conditions. Desired data for any future experiments would include time, event, placement of students, and clearly defined placement positions of students who reacted. What can be stated with some confidence for the five-student case is that each student does not rely equally on the other students for uncertainty-based information, and the five students do not coalesce as a traditional social queue based on distance from the door (calculations were done separately). Additionally, it seems that group pressure, found in a socially attractive/cohesive group, is not present in this situation. Given the results of this and previous experiments, a hypothesis is provided.

> **Hypothesis 4.1**: When in an ad hoc group, having no group-specific social norms which lead to expectation for those closer to the stimulus to react, there is little-to-no group pressure that leads to the amplification of sensation magnitude beyond what each group member feels individually.

It is important to note the words, "having no group specific social norms." The group as defined here solely exists to individually fill out a questionnaire. Helping a maintenance man who may have fallen (ambiguous) is not necessarily a social norm which requires enforcement by a group member. In western culture, this situation does require alerting staff or others who identify with group occupancy in the building and then possibly helping the maintenance man if the subject feels qualified (Sorokin's contractual society) – hence, there is further potential ambiguity.

4.6 Theft of Beer and Money (Design of Experiment Lessons Learned)

Background and Scenario: Two experiments are considered in this final analysis section. The first is an experiment by Latane and Elman (Latane and Darley, 1970, pp. 70–74, Chp. 8; Latane and Nida, 1981, pp. 311, 314–315) involving the staged theft of money. The second experiment by Latane and Darley involves the staged theft of a case of beer from a liquor store, presented in (Latane and Darley, 1969, 1970, pp. 74–77; Latane and Nida, 1981, pp. 311, 314).

In the first experiment, two conditions were evaluated (Latane and Nida, 1981, p. 314) with associated data in Table 4.4.

Condition 1: (25 trials) One naive student volunteer.
Condition 2: (16 trials) Two naive student volunteers.

TABLE 4.4 Summary Results for the Theft of Money Experiment – Data from Latane and Nida (1981, with Permission from the American Psychological Association)

Assumes all Students Observed the Event	*Total Number of Trials*	*Immediately Reported Theft*
Condition 1: Single Naive Student Observer	25	6 of 25 (0.24)
Condition 2: Two Naive (Pair) Student Observers	16	3 of 16 (0.188)

Application of the Queue Transform: Assuming all students noticed the event, for the Condition 1 single naive student who spontaneously reports the theft as the baseline case,

$$\text{Spontaneous Report: } \overline{es}_{1,0} = -ln\left(1 - \frac{6}{25}\right) = 0.2744.$$

We may now use Proposition 4.1 to obtain the theoretical probability of reporting from the Condition 2 student pairs,

$$P(R) = 1 - \left[1 - \left(1 - e^{-0.2744}\right) \cdot e^{-z(1)}\right]^2 = 1 - (1 - 0.0931)^2 = 0.178.$$

The theoretical result of 0.178 is almost identical to the empirical result of 0.188 and well within the 95 percent credibility interval of (0.066, 0.434), making it likely that all subjects observed the theft.

The second experiment involves the staged theft of a case of beer from a liquor store. This is actually a clever field experiment. Unfortunately, the experimenters did not quantify the number of paired subjects who were friends but instead made a subjective assessment while providing no justification. Finally, and as critical for our purposes, the probability of a customer reporting the theft for each of the four conditions was not provided in either reference. If this data becomes available in some useable form in the future, this would be an interesting analysis to further evaluate the reaction impact when two naive friends are the subjects.

TABLE 4.5 Summary Results from Chapter 4, Sections 4.1 through 4.6, not Including Baseline Condition Results

Chapter 4 Experiments: Reference and Condition	*Empirical Reaction Probability P (R)*	*Theoretical Reaction Probability P (R)*
Milgram (1963) – Condition 1	0.350	0.386
Milgram (1965a, 1965b) – Condition 2	0.350	0.350
Milgram (1964, 1965a) – Condition 3	0.275	0.350
Milgram (1964, 1965a) – Condition 4	0.900	0.907
Milgram (1965b) – Condition 5	0.600	0.642
Milgram (1965b) – Condition 6	0.700	0.642
Bryon and Test (1967) – Condition 2	0.029	0.027
Latane & Darley (1968) – Condition 2	0.100	0.154
Latane & Darley (1968) – Condition 3	0.380	0.394
Latane & Rodin (1969) – Condition 2	0.071	0.267
Latane & Rodin (1969) – Condition 3	0.400	0.463
Latane & Rodin (1969) – Condition 4	0.700	0.692
Clark & Word (1972) – Condition 2	0.200	0.218
Clark & Word (1972) – Condition 3 (Assuming Two Independent Groups)	0.400	0.354
Latane & Nida (1981) – Condition 2	0.188	0.178

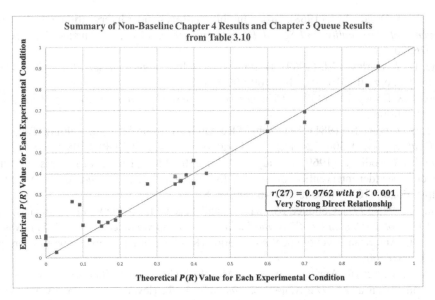

FIGURE 4.3   Empirical Results Compared to Theoretical Results for Queue and Non-Queue Social System Examples.

4.7 Cumulative Summary of Queue Transform Results

Table 4.5 provides summary results for experiment conditions analyzed in Sections 4.1 to 4.6, excluding baseline condition results used for initializing the queue transform for the given experiment. Figure 4.3 includes data from Tables 3.9 and 4.5 with a resulting Pearson's $r(27)$ value of 0.9762 with $p < 0.001$.

References

Bryon, J., & Test, M. (1967). Models and Helping: Naturalistic Studies in Aiding Behavior. *Journal of Personality and Social Psychology, 6*(4), 400–407. https://doi.org/10.1037/h0024826

California Driver Handbook. (2020). *State of California: Department of Motor Vehicles.* Retrieved from https://DMV.CA.gov

Clark, R., & Word, L. (1972). Why Don't Bystanders Help? Because of Ambiguity? *Journal of Personality and Social Psychology, 24*(3), 392–400. https://doi.org/10.1037/h0033717

Dehaene, S., Izard, V., Spelke, E., & Pica, P. (2008). Log or Linear? Distinct Intuitions of the Number Scale in Western and Amazonian Cultures. *Science, 320,* 1217–1220. https://doi.org/10.1126/science.1156540

Festinger, L. (1957). *A Theory of Cognitive Dissonance.* Stanford University Press. Retrieved from www.sup.org/books/title/?id=3850

Hubler, A., & Pines, D. (1994). Prediction and Adaptation in an Evolving Chaotic Environment. In G. Cowan, D. Pines, & D. Meltzer (Eds.), *Complexity: Metaphors, Models, and Reality* (pp. 343–382). Proceedings Volume XIX, Santa Fe Institute Studies in the Sciences of Complexity, Addison-Wesley Publishing Co.

Klir, G. (2006). *Uncertainty and Information: Foundations of Generalized Information Theory.* John Wiley & Sons, Inc. https://doi.org/10.1002/0471755575

Latane, B., & Darley, J.M. (1968). Group Inhibition of Bystander Intervention in Emergencies. *Journal of Personality and Social Psychology*, *10*(3), 215–221. https://doi.org/10.1037/h0026570

Latane, B., & Darley, J.M. (1969). Bystander "Apathy". *American Scientist*, *5*(2), 244–268. https://doi.org/10.1016/0022-1031(69)90046-8

Latane, B., & Darley, J.M. (1970). *The Unresponsive Bystander: Why Doesn't He Help?* Prentice Hall Publishing.

Latane, B., & Nida, S. (1981). Ten Years of Research on Group Size and Helping. *Psychological Bulletin*, *89*(2), 308–324. https://doi.org/10.1037/0033-2909.89.2.308

Latane, B., & Rodin, J. (1969). A Lady in Distress: Inhibiting Effects of Friends and Strangers on Bystander Intervention. *Journal of Experimental Social Psychology*, *5*(2), 189–202. https://doi.org/10.1016/0022-1031(69)90046-8

Lewin, K. (1945/1997). Conduct, Knowledge, and Acceptance of New Values. In G.W. Lewin (Ed.), *Resolving Social Conflicts and Field Theory in Social Science* (pp. 47–57). Washington, DC: American Psychological Association (Kindle Edition). https://www.apa.org/pubs/books/4318600

McFadden, R. (2016, April 5). Winston Moseley, Unsparing Killer of Kitty Genovese, Dies in Prison at 81. *The New York Times*, 21A. Retrieved from https://www.nytimes.com/2016/04/05/nyregion/winston-moseley-81-killer-of-kitty-genovese-dies-in-prison.html

Milgram, S. (1963). Behavioral Study of Obedience. *Journal of Abnormal and Social Psychology*, *67*(4), 371–378. https://doi.org/10.1037/h0040525

Milgram, S. (1964). Group Pressure and Action Against a Person. *Journal of Abnormal and Social Psychology*, *69*(2), 137–143. https://doi.org/10.1037/h0047759

Milgram, S. (1965a). Liberating Effects of Group Pressure. *Journal of Personality and Social Psychology*, *1*(2), 127–134. https://doi.org/10.1037/h0021650

Milgram, S. (1965b). Some Conditions of Obedience and Disobedience to Authority. *Human Relations*, *18*(1), 57–76. https://doi.org/10.1177/001872676501800105

Milgram, S. (1970). The Experience of Living in Cities. *Science*, *167*(3924), 1461–1468, https://doi.org/10.1126/science.167.3924.1461

Ofgang, E. (2018, May 22). Revisiting the Milgram Obedience Experiment Conducted at Yale. *New Haven Register*. Retrieved from www.nhregister.com/news/article/Revisiting-the-Milgram-Obedience-Experiment-12870042.php

Parkinson, C., Kleinbaum, A., & Wheatley, T. (2018). Similar Neural Responses Predict Friendship. *Nature Communications*, *9*(332), 1–14. https://doi.org/10.1038/s41467-017-02722-7

Patton, S. (1977). Milgram's Shocking Experiments. *Philosophy*, *52*(202), 425–440. https://doi.org/10.1017/S0031819100028916

Scanning for Hazards. (2022). *My License: The Hazard Perception Test*. Government of South Australia: Department of Infrastructure and Transport. Retrieved from https://mylicence.sa.gov.au/the-hazard-perception-test/scanning

Tajfel, H., & Turner, J.C. (2004). The Social Identity Theory of Intergroup Behavior. In J.T. Jost & J. Sidanius (Eds.), *Political Psychology: Key Readings* (pp. 276–293). Psychology Press. https://doi.org/10.4324/9780203505984-16

Zipf, G. (1948). *Human Behavior and the Principle of Least Effort: An Introduction to Human Ecology*. Ravenio Books (Kindle Edition). https://doi.org/10.1002/1097-4679(195007)6:3<306::AID-JCLP2270060331>3.0.CO;2-7

Zoccola, P., Green, M., Karoutsos, E., Katona, S., & Sabini, J. (2011). The Embarrassed Bystander: Embarrassability and the Inhibition of Helping. *Personality and Individual Differences*, *51*, 925–929. https://doi.org/10.1016/j.paid.2011.07.026

5

FROM THE QUEUE TO THE COMMONS

Thus, the queue seems to be a social system in which position must be earned, and one can only earn it through waiting.

R. J. Toledo, May 29, 1978, Stanley Milgram Papers (MS 1406, Box 101, Folder 138 with permission from the Estate of Alexandra Milgram)

High status or other disproportionate benefits must be earned.

Wilson et al. (2013, p. S22, with permission from Elsevier Publishing)

Direct consequences for this work so far have been to establish a relationship between Fechner's Law and the power law and to provide the ability to accurately model western queues and propose a rough modeling approach for some basic queue-like social system experiments within the western culture. These are academically significant in themselves, but otherwise are just parlor tricks if there is no intention of applying this work to something more meaningful. To this end, and as a starting point, let's consider the queue as a common-pool resource system (Ostrom, 1990, p. 90) and apply the queue transform to quantitatively model the utility of certain design principles and how they apply in maintaining stability of the queue. To fall into the category of a CPR, one or more resource units offered in a queue must be subtractable (rival good) yet available to all in the community of intertest (non-excludable). If a queue can be shown to represent a common-pool resource system, then the queue transform may be applied to operationally demonstrate the importance of certain design principles necessary to facilitate social stability in the queue. Consequently, it may then provide a window for looking into the larger social system in which the queue as a social system resides. Applying and building on the psychophysics-based queue transform to analyze more complex western social systems using core design principles for social stability as a starting point might be useful in beginning to address the vision Elinor Ostrom (2012, pp. 68–71) had of creating a multidisciplinary multiple-tier framework in which to analyze social–ecological systems.

DOI: 10.4324/9781003325161-6

5.1 The Common-Pool Resource

The focus of this chapter is on types of goods and services necessary for meeting our day-to-day and long-term social needs. In this context, goods and services that meet our social needs will be defined as social resources. These social resources in turn extend from rival to non-rival social resources, which in turn extend from excludable to non-excludable (Fisher et al., 2009, p. 647). If a resource is rival, then if part of the resource is used, less exists for use by others in the community. If a resource is excludable, then its use may be restricted to certain individuals in the community. Elinor Ostrom (2005) indicates four basic types of resources: *toll goods* (predominantly non-rival and excludable), *public goods* (predominantly non-rival and non-excludable), *private goods* (predominantly rival and excludable), and *common-pool resources* (predominantly rival and non-excludable). One further level of distinction is that within each of these categories there are two components – the resource system and the resource unit (Ostrom, 1990, p. 30). It is the resource system that allows access to the desired resource unit, such as a parking garage allowing access to a finite number of parking spaces and spaces closer to the exit being more desirable. Barkin and Rashchupkina (2017) discuss the gray areas which can exist when distinguishing between these four categories, and how a category may change based on the way it is viewed. It may be that one particular category shares some similarity to one or more of the remaining three, or that a complete shift in category may occur based on what resource unit is being considered. A definition of common-pool resource, seemingly to have been settled on beginning with Ostrom et al. (1994), and as used more recently in Ostrom (2002, p. 1317), is that common-pool resources are systems having resource units that renew at a limited rate so that one person's use subtracts from the number of resource units available to others in the system. With the basic components and definition identified, let's now consider the train ticket queue in terms of a common-pool resource.

To frame the train ticket queue as a common-pool resource in a socially representative manner, let's consider two hypothetical and geographically separate communities each relying on their single train that leaves once a day. In these two communities, the community member who obtains a seat on a train using the queue as a resource system is equated with obtaining a means for incrementally improving the member's long-term prosperity. Let the first community have what we will label an ideal queue. In the ideal queue, train tickets are affordable by everyone in the community. Tickets at this community's ticket booth begin to be sold to those in the first-come first-served queue 20 minutes before each daily train departure. At the end of the twenty-minute period, sales are discontinued for the day, and each member's position is adjusted for the next day based on what benefit to the community each provided that day or has shown the potential to provide. In the second community, there are two social subgroups – a higher-class subgroup and a lower-class subgroup. The higher-class subgroup has the legal right to intrude at the front of the queue without any social repercussions, while those in the lower-class subgroup follow the rules of the first community but may or may not be able to afford the ticket depending on circumstances at the time – if and when they ever reach the ticket server. If unable to afford the ticket after having reached the ticket server, the lower-class member in the second community must return to the end of the queue and start over again. In these two hypothetical communities, time in queue and ability to obtain a ticket are the two interconnected social resource units of interest.

If a member in either community earns higher status and thus position in the queue (i.e., greater prestige) through providing above normal benefit or promise of benefit to the community (Cheng, 2020, p. 238; Ostrom, 1990, p. 96), then those displaced behind him or her consider this a legitimate intrusion and appreciate the benefit the community member provides, thereby reducing their dissonance. In the second community, if a dominant higher-class member intrudes (enters) at the front of the queue, then all lower-class community members lose time and must wait a little longer based on service time at the ticket server and any subsequent intrusions that may occur by other higher-status members. Work related to this type system indicate that the second community's social resource system has reduced the social stability compared to the first (Cheng, 2020, p. 239; Wilson, 2015).

As a simple quantitative example demonstrating the impact of intrusion, consider a queue whose single server has an exponentially distributed mean service time of two minutes and that twenty community members are waiting in queue to obtain a train ticket for a train that leaves in twenty minutes. Using the Erlang Type k density function (Gross and Harris, 1985, pp. 171–172), the probability that the queue member in the kth position gets his or her ticket before the train leaves, designated $P_k(ticket)$ is

$$P_k(ticket) = \int_0^{20min} \frac{(.5/min)^k}{(k-1)!} \cdot (x)^{k-1} \cdot e^{-(0.5/min) \cdot x} dx$$

Therefore, the probability the first queue member obtains his or her ticket before the train leaves is almost guaranteed at $P_1(ticket) = 0.99995$. The probability of the second member getting a ticket before the train leaves is $P_2(ticket) = 0.99947$. The probability of the tenth member getting a ticket before the train leaves is $P_{10}(ticket) = 0.5358$, and the probability of the twentieth member getting a ticket before the train leaves is $P_{20}(ticket) = 0.0033$. If, while waiting in line, an intrusion occurs, then the probability of obtaining a ticket is decreased for all members behind the intrusion point. For example, when an intrusion occurs at the front of the queue, the tenth member now becomes the eleventh member having probability $P_{11}(ticket) = 0.4107$. Hence, by just one intruder taking a position ahead of the tenth member, the probability of the now former tenth member obtaining a ticket has been significantly reduced. Basically, time in queue and queue position (status) are interchangeable (Zhou and Soman, 2003, p. 521). The higher the queue status, the less time in queue and the more likely that a ticket is obtained that day leading to greater opportunity.

In general, if members had been waiting for an ice cream cone, the members behind the intrusion point would lose time as a resource unit but still have access to an ice cream cone. If an intrusion occurs while waiting for the train ticket, the probability of a member obtaining a ticket along with the opportunity that the ticket provides is reduced. If the member fails to get a ticket, he or she loses both time and opportunity that day. There are many similarities between this example and day-to-day life, with history as data providing this theme repeatedly. We are born with certain gifts or capabilities that could benefit the community if the desire exists to apply them, given the opportunity exists to employ them. Ideally, we work to improve our status in the queue through benefitting the community and waiting in an effort to increase the probability of obtaining the desired ticket and opportunity it affords. If we miss the chance, we try again. Others though, for whatever reason, may either intrude near the front of the queue or walk to the front of the queue due to inherited status – status that may or may not have been earned. The intruder or the higher-status member who offers no benefit to the community may then be considered a free-rider. Time and the ticket to greater opportunity are then woven together in this manner, so as a community member, the objective is to make the resource system equitable, meaning each member earns his or her position in queue that day through merit or potential benefit to the community based on fair and quantifiable metrics.

In the first community, time and tickets as a function of time are rival non-excludable resource units. In the second community, time is rival and non-excludable, but as a function of time tickets may be viewed as rival but excludable. Hence, in all or part, the queue resource system falls under the definition of a common-pool resource. Elinor Ostrom (1990, pp. 69–75, 90, 136) identified eight design principles that a social community/group must meet to support equity and optimize the chance for resource system stability within the social group sharing the CPR. Traditional CPR examples are fish harvested from a specific lake which is shared by a community, community-owned land used for grazing of its cattle, a community's use of its forest for wood, or irrigation communities relying on an equitable distribution of water. The eight design principles, identified by Ostrom, have since been further analyzed and modified (Wilson et al., 2013, p. S22). In summary, for community members to increase the stability of their resource system within the context of this work while maintaining the traditional structure presented in Wilson et al. (2013, p. S22, With permission from Elsevier Publishing), they should:

Design Principle 1) **Maintain and clearly define resource system boundaries**.

Design Principle 2) **Require that appropriation of rival resource units be commensurate with benefits each member provides to the resource system and community as a whole (i.e., status must be earned)**.

Design Principle 3) **Ensure that resource system members are able to make and modify rules (social norms) through consensus**.

Design Principle 4) **To avoid a tragedy of the commons (Hardin, 1968), establish the capability to efficiently monitor the rival resource units against free-riding and exploitation within the community relying on the resource system**.

Design Principle 5) **Have a means of applying community-accepted graduated sanctions against free-riding and other appropriator violations**.

Design Principle 6) **Establish conflict resolution mechanisms within the resource system that are efficient and acceptable to the community using it**.

Design Principle 7) **Have the right to modify the resource system structure to accommodate local and/or changing circumstances**.

Design Principle 8) **Establish constructive (effective) coordination with other critically linked resource systems that are necessary for the proper operation of the community's resource system**.

Cox et al. (2010) provide a thorough independent meta-analysis of empirical studies following the governance process proposed by Ostrom and, in doing so, identify design principles from these eight having the greatest positive impact on resource system stability. Three of the design principles, principles 1, 2, and 4, rank highest in maintaining resource system stability based on the metrics used. The remainder, though important, seem to address external coordination and internal rules necessary to support these same three design principles. Additionally, the effectiveness of design principle 4 is dependent on design principle 1 (recall the information booth queue of Chapter 3). Therefore, subsequent focus of quantitative analysis of design principles will for now be placed on design principles 2 and 4.

Though based on CPRs such as irrigation systems, communal forests, inshore fisheries, and grazing land (Ostrom, 1990, pp. 20–21), it is argued by Wilson et al. (2013) that these principles may have a wider application for addressing nearly any situation where members of a nested CPR must cooperate and coordinate to achieve common goals within a larger and necessarily interactive horizontal and/or vertical social system. The queue as a CPR is one such nested, albeit very basic social system.

Since position in queue relates directly to mean waiting time in queue and probability of obtaining a ticket, time in queue will be the only resource unit to be addressed. Though the ticket being sought through use of the queue may be viewed as a private good where payment for the good excludes its use from those who do not or cannot pay, it is time in queue that allows access to the good being sought which in itself a subtractable resource unit since every human lives for a finite amount of time. Time in queue, waiting, is another form of payment and means to gain status toward increasing the probability of reaching the desired train ticket. But a community member will wait only so long before leaving or reacting when the resource system is perceived to provide an unfair and unequal distribution of resources (Tajfel and Turner, 2001, pp. 98–99, 2004, p. 287). If that is the situation, access to the ticket has become unreliable or nonexistent, and the queue as a resource system will collapse for the lower-class members in the second community if too many higher-class community members exist within that same community.

5.2 The Queue as an Open-Access Common-Pool Resource

The queue as a basic social system is a continuum of the larger social community of which the queue members belong, sharing certain social norms important to and accepted by the community. Members do not leave their community or its norms when entering the queue but instead adjust their identity as pedestrians, motorists, parents, students, or some other social group member with specific social norms to that of a queue member adhering to a subset of social norms and identity consistent with that of the larger community in which they reside. Hence, since it has been shown that the social queue can be reliably modeled for the given data, and if it may be shown to meet the eight design principles, then larger more complex social communities may eventually be modeled, at least in part, and analyzed for stability.

In Ostrom (1990, p. 48), the definition of open-access CPR is carefully laid out. Let us begin by equating the parlance of the open-access CPR environment with that of the western queue environment. Resource units at a train station are waiting time in queue, or equivalently probability of obtaining a ticket based on queue status. Appropriators of the resource units in our example are the community queue members waiting for the tickets they wish to obtain from the ticket server. Provisioning by queue members comes in the form of time in queue, monitoring and maintenance of queue integrity, and monetary payment (always affordable or not always affordable) to the ticket server by the queue member. Let's now look at the queue as a CPR from the quantitative standpoint to better understand the implications of the eight design principles and how the two hypothetical communities discussed may relate to larger more complex social systems.

5.2.1 Design Principle 1 – Clearly Defined Resource System Boundaries

A strong group identity helps create social boundaries. How strongly we identify with a social group may best be measured by the strength of attraction we have for being part of the group versus for individuals in the group (Stangor, 2016, p. 24). In other words, if joining the group provides a benefit to the individual, then it is attractive. Whether the perceived benefit is from increased self-esteem or the ability to obtain a desired resource that would otherwise be more difficult to obtain, individual benefit is gained by joining the group as a member. Once an individual enters the western social queue for a train ticket, he or she may now obtain the desired train ticket assuming sufficient time and tickets are available. In doing so, he or she now identifies as a queue member who should adhere to the associated social norms and provisioning efforts necessary to maintain queue structure and stability. If the desired benefit is not realized by the member, the member then must make a choice through comparison of existing and available alternative resource systems while accounting for risks in remaining versus transitioning to any alternative (i.e., emigration to another queue as an example).

Boundaries of the resource system could refer to geographic boundaries, physical boundaries, or social boundaries. The physical and social boundary of a queue is based on whether you are a waiting member of the queue, defined by being in a hierarchical line of members, or whether you are outside of the queue and therefore not a member. Being in queue implies you are part of a social group (i.e., community) having a defined and understood social order for reaching the server and obtaining the desired resource (Schmitt et al., 1992, p. 815; Kuzu, 2015). In the ideal western first-come first-served queue, assuming no intrusions or higher-class members are able to enter at the front of the queue, the community member with the highest order in queue is served next, and the person with the lowest order in queue is served last from the set of currently existing community members.

Expected social queue behavior (western social norms) implemented to maintain the social and physical boundary of the queue and its subsequent group attraction include:

1. Entrance into the queue from the community having access to it is at the end of the existing queue member order.
2. Queue members are expected to maintain integrity of the group by maintaining a clearly defined boundary between queue group members and nonmembers. This is managed by maintaining close proximity and a linear hierarchical order.
3. Queue members expend time to obtain a specific resource. If a member is not in queue for that resource, then the member does not belong in the queue and should not obtain the resource.

Based on this, when properly maintained, the queue resource system has a clearly defined physical and social boundary.

5.2.2 Design Principle 2 – Proportional Appropriation and Provision

For the queue resource system example, appropriation comes in the form of obtaining the desired train ticket from the ticket server. Provisioning (effort) comes in the form of monetary payment to the ticket server for the ticket resource unit sought, expending time resource units waiting in queue, and any additional individual member effort expended in maintaining queue group social norms. The latter is indicated since additional external security to maintain queue integrity ultimately comes at a cost to the queue members through an increase in ticket prices. In other words, if you want the ticket at the existing low price, you must pay for it in three different ways (i.e., payment, waiting, and maintenance of queue integrity). If community members or subgroup members are unable to obtain a ticket in reasonable time, or if social norms of the queue are unable to be maintained, then attractiveness of being a member of that queue resource is reduced, possibly, to the point of completely losing its attractiveness to affected queue group members.

Queue benefits versus provisions lead to attractiveness or unattractiveness of the queue. This may seem rather obvious, but it gets to the heart of many problems that occur in the real world. If a ticket is excludable, and if there is only one train that supports each specific community, then whoever controls the price of the train ticket may increase their price until it becomes more beneficial to higher class members, leading lower-class community members to find an alternate means of obtaining the desired ticket – if one exists. Overcharging may be viewed as a form of free riding, unnecessarily increasing the ticket price for greater profit without providing commensurate benefit to the community, while restricting tickets to only those who can afford to pay for them.

The second form of provisioning is that of waiting in queue and maintaining your relative position. To leave the queue and expect to return to the same relative position are not acceptable social norms unless they are agreed beforehand with the queue members behind. This is a matter of equity. If the other members need to wait in line, then the member wanting to leave and return is implying he or she has better things to do and should not have to waste time waiting but instead return when it is his or her turn to be served. On the other hand, if this were a common practice, then queue boundaries would not exist, and those entering the queue would not know who was ahead of them or how long the wait might be. The latter situation in western cultures could easily end in numerous emotional reactions.

The third and final form of provisioning is maintaining integrity of the queue. By forming a queue resource system, the boundary is established by each member getting behind the next in order of arrival. This approach makes it practical for queue members to determine when someone is trying to intrude into the queue by entering it at some point other than the very end. In the event that an intruder appears and tries to cut into the queue ahead of others already waiting, it is up to the members at the point of the intrusion, if counter to existing queue resource system social norms, to react and either expel the intruder or at least embarrass him or her to exact a cost for their not abiding by the social norm. This does come at a potential cost to the queue member(s) who react, in the form of stress if confrontation occurs, potential embarrassment if misinterpretation of the situation occurred, or physical harm at the extreme.

The following comments are made to clarify the role and need for members to expend effort (provision) in defending the queue from intrusion (free riding) when necessary and for minimizing delays that may impact those behind by subtracting from their time resource units.

1. Once in the queue, you do not move from your position in the social queue group to displace the position of others ahead of you. If no plausible explanation or community benefit is understood to warrant the change in position, then this may indicate many things to the remaining queue members, including social inequity – that the violator believes without justification that his or her time is more important than those he or she has jumped ahead of.
2. Defending against intrusion is a provision expected of queue members when the intrusion event occurs. This is emotionally uncomfortable to the defending queue member but is necessary to maintain integrity of the queue and to avoid loss of subtractable resources (time/position in queue).
3. When service is complete for the current member being served, the member at the top of the queue order (next in line) immediately steps up to be served by the server. Delay in doing so, such as not preparing ahead of time for a quick transaction, is not appreciated and wastes time resource units of the remaining queue members.

Hence, there is a process and expectation that benefits of appropriation are proportional to the costs of provision.

5.2.3 Design Principle 3 – Modify Rules Through Consensus

In maintaining the social norms and boundary of the social queue, members may on occasion have to modify rules or create rules for special circumstances. Since maintaining attractiveness of the social queue is paramount to members, they should have a say in how rules may be modified. Helweg-Larson and LoMonaco (2008, pp. 2380, 2388–2390) provide an example where queue members waiting in queue for U2 rock concert tickets generally preferred to make rule modifications on their own. Doing so maintains group identity through group member participation and results in maintaining attractiveness of the group to the group members. In some situations, it was noted that queue members would prefer third-party enforcement to act for them, thereby freeing the queue members from embarrassing and possibly emotionally difficult situations of confrontation with an intruder (see Axiom 2 from Chapter 1). In some manner though, this third-party security comes at a cost (money, reduced effectiveness, etc.), and members of the group must be able to support that cost and also see it is beneficial to them.

Hence, when modification to the rules is deemed necessary by queue group members, members in queue do have the option for collective-choice arrangements.

5.2.4 Design Principles 4 and 5 – Monitoring Access to the Resource and Effective Sanctions Against Violators

The social queue provides a low-cost (measured in effort) approach to monitoring access to the resource via the queue. In western queues, the default social norm has traditionally been first-come, first-served. To clarify how this could be violated and how effectively it may be monitored, consider the example where someone not in the queue enters the queue at the front (next to be served). That person has not waited, nor has he or she helped to maintain the social norms of the queue group while waiting. In effect, the intruder is an outsider who has not provided the necessary provision to the queue group and therefore falls into the category of a free-rider. Free-riders are those that benefit from the group's resource but do not help in maintaining that resource. As more free-riders enter the group, the attractiveness of the group decreases for those members who support the norms of the group and maintenance of its social boundaries.

To quantify the train ticket queue example, consider queue group members who need a ticket for a train leaving in twenty minutes. The existing members have been waiting in line, which is a cost to them. They are maintaining the social norms of the group which entails confronting intruders and through the social cost of self-restraint by not jumping ahead of others. If we assume a simple queue for our purposes, where the single server has a Poisson distributed service rate (μ) and the arrival rate (λ_2) of members into a single server queue with their interarrival time exponentially distributed (i.e., resulting in an M/M/1 queue), then the mean waiting time (W_{q2}), including member time in service, for the members incurring the social cost of the first-come first-served queue is (Gross and Harris, 1985, p. 77):

$$W_{q2} = \frac{1}{\mu - \lambda_2}$$

If the mean service rate is $\mu = 60$ *customers per hour*, and the mean interarrival time of members into the queue is every 90 seconds, then $\lambda_2 = \dfrac{60 \text{ minutes}}{1.5 \text{ minutes}} = 40$ customers per hour. This results in a mean waiting time of

$$W_q = \frac{1}{(60-40)/hour} = \frac{1}{20} \; of \; an \; hour = 3 \, minutes.$$

Now consider what happens if intruders enter non-preemptively at the front of the queue, behind the member currently being served, with arrival rate λ_1. In this situation, where the intruders might represent higher status (higher priority) individuals with head of the queue privileges, then everyone waiting in line behind the front queue position will have to wait longer. Consider what happens when $\lambda_1 = 5$ high-priority "intruders" enter every hour. Then (Gross and Harris, 1985, p. 199)

$$letting \; \lambda = \lambda_1 + \lambda_2 = 40 + 5 = 45,$$

the higher status "intruders" have a mean waiting time of,

$$W_{q1} = \frac{\lambda}{\mu \cdot (\mu - \lambda_1)} = \frac{45}{60 \cdot (60-5)} = 0.82 \, minutes.$$

Lower status queue members though now have a mean waiting time of

$$W_{q2} = \frac{\lambda}{(\mu - \lambda) \cdot (\mu - \lambda_1)} = \frac{45}{(60-45) \cdot (60-5)} = 3.3 \, minutes.$$

Some people learn more quickly than others, and some care more about social norms than others, so as in law enforcement, social enforcement should allow for graduated sanctions by queue members based on an intruder's age, nature of offense, and frequency of offense. As observed in Milgram et al. (1986), and displayed in the lower half of Table 3.2, sanctions varied from no reaction to physical ejection of the intruder based on the situation in which the offense took place. With regard to increasing sanctions based on frequency of offense, Oberholzer-Gee (2006) demonstrated in a field experiment that an intruder buying his way into the queue was allowed the first time but firmly rejected the second to the point where after 15 trials the experiment was stopped as it was deemed to be too physically dangerous. Using this data, the probability of more severe sanctions increases with the importance of the queue resource (i.e., nature of offense) or frequency of the offense. From an importance standpoint, arguing in a similar manner as Helweg-Larson and LoMonaco (2008, p. 2389), those queue members waiting for a train ticket likely place greater importance on that resource than those members waiting in queue for their turn for a resource provided by a bank or token booth having basically unlimited resources and negligible time constraints (unless in queue just before closing – this would be a good experiment for the brave).

Define a free-rider here as one who does not expend effort to maintain social norms or time in queue as other norm-abiding members do to gain access to the resource, but still expects access to the resource (Ostrom, 2005, pp. 24, 79–80, 262). As the number of successful intrusions by free-riders increases per hour, the attractiveness of the queue diminishes for those group members who wait in queue and try to enforce the first-come first-served western social norm. If the social norms become unenforceable, then referring to Design Principle 1 and 2, it is likely that the norm-abiding queue members will try to alter the resource system to their advantage. In the worst-case scenario, if Design Principles 4 and 5 were to fail, queue members would no longer find the queue beneficial to them and would either leave or revolt. This result may be referred to as a tragedy of the commons.

Hence, low-cost monitoring of access into the queue is possible, but adequate sanctions must be enforceable by queue members or their hired surrogates, as monitors, for the purpose of enforcing membership requirements and maintaining attractiveness of the queue.

5.2.5 Design Principle 6 – Conflict Resolution Mechanisms

Addressing Design Principle 6 for a queue environment offers an interesting interpretation of queue member reaction as a function of position relative to the intrusion point. It has already been shown how the probability of member reaction decreases as number of members $|UI|$ between the intrusion point and position of interest increases. As the queue transform indicates, reactions by each of the $|UI|$ queue members ahead of the position of interest reduce the probability of reaction at the position of interest. From this, it may be interpreted that reactions are the means of internal conflict resolution addressing the social norm deviation caused by the intrusion. Since typically, the resource the queue member is waiting for is not required for survival, if no reaction occurs, resolution may come about through queue members accepting the intrusion as acceptable for some reason. If reaction does occur, then resolution may come about through queue members appreciating that the intruder was appropriately embarrassed or ejected by those queue members who reacted, who were close by, and who best understood the extent of the infraction. If the importance of the queue's resource being sought is in fact important to survival or well-being, then expect more violent member reactions where resolution relies with increasing probability on social group (queue) expulsion and/or personal injury (Zipf, 1948, loc 10850–10916; Helweg-Larson & LoMonaco, 2008, p. 2389).

Hence, for a queue, localized member reaction to an apparent social norm deviation and the resultant uncertainty information provided to the other members are a form of conflict resolution for the queue community. If there is conflict between queue members, typical experience indicates that the members closest to the situation will assist in resolving the conflict through providing information that substantiates one or the other's claim.

5.2.6 Design Principle 7 – Right to Modify Resource System Structure

A first-come first-served queue organizes under conditions that we take for granted, given that its basic nature and that its formation for various resources are common in the western culture. When equity of resource distribution is expected based on waiting time, the queue has been the means for citizens in western nations to obtain the desired resource while maintaining the western expectation of equality between citizens. The queue as a system has been modified in some cases over time based on circumstance or as a means to improve efficiency – one example being where each new member takes a ticket when entering a resource system thereby alleviating the requirement for a physical queue. This in turn alters enforcement responsibility to the server, where only the next member in queue is served based on the member having a ticket with the number next to be served.

Hence, in western communities, the queue resource system can be modified and accepted by queue members under circumstances that benefit the community members using the queue.

5.2.7 Design Principle 8 – Effective Coordination Between Interconnected Resource Systems

What must be addressed first is how the queue fits into and interacts with a larger social system. To do this, consider the queue for train tickets and the train company which is providing the desired transportation resource for queue members who obtain a ticket. If there were no train, there would be no reason for the ticket queue to exist. The queue is a social group (Mann, 1969), and the train company could be considered a social group. Combined, both create a larger and interrelated social system. Coordination for selling tickets at the server must occur with the train company since selling unlimited tickets for a limited number of seats would cause emotional discomfort to the customer who bought a ticket but was unable to get his or her train that day as desired.

Hence, the train company and the server of the queue selling the ticket resource for the train transportation resource have first-order relevance to one another and must coordinate, otherwise, one or the other or both are no longer attractive to the queue group members. There are obviously many additional interrelationships that are required to support feasibility and operation of this resource system, but the train company and ticket resource system are sufficient to clarify the concept.

5.3 Mapping Complex Social Commons to the Western Queue Social Commons

We have observed that experiments discussed in Chapter 4 demonstrate social systems with underlying queue-like structures, which in most cases (when the necessary experiment details are available) may be transformed/mapped in a manner that allows the use of the queue transform. Likewise, we have just shown that the western social queue meets the eight design principles necessary for maintaining a stable commons environment. The next step is to use the queue transform for mapping appropriate parts of more complex queue-like social resource systems into the queue environment. One such resource system is the irrigation canal commons which is widely written about from many parts of the world. The analysis of such a social system goes beyond the intended effort of this book, but in evaluating the South African irrigation community and its external support system in Dzindi, for instance, it becomes readily apparent why the commons has been failing over the last two decades. Underlying culture must be accounted for, and no social resource system is an island in itself (Design Principle 8) but instead is part of a larger interacting whole. If the parts act in a coherent supportive process, then each survives and thrives. If there is disharmony between certain parts of the whole, then the overall system will flounder to the extent that social disharmony is experienced as a result.

It is proposed that the last five chapters have possibly begun to address the vision of Elinor Ostrom as she stated it in (Ostrom, 2012, pp. 69–70) and as introduced in the first chapter. With that done, it remains to define an algebraic social space as qualitatively alluded to by Lewin (1997) and Sorokin (1959). An algebraic social space allows for the construction of more complex theorems and supporting analytical structures, which may in turn improve opportunities for reliable, testable, and repeatable quantitative analysis of larger social systems. The ultimate goal is for a systematic means toward better understanding the importance of certain social variables and from that how to improve the social stability of our commons. With an algebraic social space defined at the group level, we will then finish with an historical example of how what has been learned may be applied.

References

Barkin, S., & Rashchupkina, Y. (2017). Public Goods, Common Pool Resources, and International Law. *The American Journal of International Law, 111*(2), 376–394. https://doi.org/10.1017/ajil.2017.9

Cheng, J. (2020). Dominance, Prestige, and the Role of Leveling in Human Social Hierarchy and Equality. *Current Opinion in Psychology, 33*, 238–244. https://doi.org/10.1016/j.copsyc.2019.10.004

Cox, M., Arnold, G., & Tomas, S. (2010). A Review of Design Principles for Community-Based Natural Resource Management. *Ecology and Society, 15*(4), 38. Retrieved from www.ecologyandsociety.org/vol15/iss4/art38/

Fisher, B., Turner, R., & Morling, P. (2009). Defining and Classifying Ecosystem Services for Decision Making. *Ecological Economics, 68*, 643–653. https://doi.org/10.1016/j.ecolecon.2008.09.014

Gross, H., & Harris, C. (1985). *Fundamentals of Queueing Theory* (2nd ed.). John Wiley and Sons.

Hardin, G. (1968). The Tragedy of the Commons. *Science, 162*, 1243–1248. Retrieved from www.jstor.org/stable/1724745

Helweg-Larson, M., & LoMonaco, B. (2008). Queuing among U2 Fans: Reactions to Social Norm Violations. *Journal of Applied Social Psychology, 38*(9), 2378–2393. https://doi.org/10.1111/j.1559-1816.2008.00396.x

Kuzu, K. (2015). Comparisons of Perceptions and Behavior in Ticket Queues and Physical Queues. *Service Science-Institute for Operations Research and the Management Sciences (INFORMS), 7*(4), 294–314. https://doi.org/10.1287/serv.2015.0116

Lewin, K. (1997). *Resolving Social Conflicts and Field Theory in Social Science* (G.W. Lewin, Ed.). American Psychological Association (Kindle Edition). Retrieved from www.apa.org/pubs/books/4318600

Mann, L. (1969). The Waiting Line as a Social System. *American Journal of Sociology, 75*(3), 340–354. https://doi.org/10.1086/224787

Milgram, S., Liberty, J., Toledo, R., & Wackenhut, J. (1986). Response to Intrusion into Waiting Lines. *Journal of Personality and Social Psychology, 51*(4), 683–689. https://doi.org/10.1037/0022-3514.51.4.683

Oberholzer-Gee, F. (2006). A Market for Time Fairness and Efficiency in Waiting Lines. *KYKLOS, 59*(3), 427–440. https://doi.org/10.1111/j.1467-6435.2006.00340.x

Ostrom, E. (1990). *Governing the Commons*. Cambridge University Press. https://doi.org/10.1017/CBO9781316423936

Ostrom, E. (2002). Common-Pool Resources and Institutions: Toward a Revised Theory. In B. Gardner & G. Rausser (Eds.), *Handbook of Agricultural Economics – Vol 2A Agriculture and Its External Linkages*. North-Holland Elsevier Science B.V.

Ostrom, E. (2005). *Understanding Institutional Diversity*. Princeton University Press (Kindle Edition).

Ostrom, E. (2012). *The Future of the Commons*. The Institute for Economic Affairs. Retrieved from http://ssrn.com/abstract=2267381

Ostrom, E., Gardner, R., & Walker, J. (1994). *Rules, Games, and Common-Pool Resources*. University of Michigan Press. https://doi.org/10.3998/mpub.9739

Schmitt, B., Dube, L., & Leclerc, F. (1992). Intrusions Into Waiting Lines: Does the Queue Constitute a Social System? *Journal of Personality and Social Psychology*, *63*(5), 806–815. https://doi.org/10.1037/0022-3514.63.5.806

Sorokin, P. (1959). *Social and Cultural Mobility*. The Free Press. Retrieved from https://ia801604.us.archive.org/8/items/in.ernet.dli.2015.275737/2015.275737.Social-And_text.pdf

Stangor, C. (2016). *Social Groups in action and Interaction*. Routledge. https://doi.org/10.4324/9781315677163

Tajfel, H., & Turner, J.C. (2001). An Integrative Theory of Intergroup Conflict. In M.A. Hogg & D. Abrams (Eds.), *Intergroup Relations: Essential Readings* (pp. 94–109). Psychology Press.

Tajfel, H., & Turner, J.C. (2004). The Social Identity Theory of Intergroup Behavior. In J.T. Jost & J. Sidanius (Eds.), *Political Psychology: Key Readings* (pp. 276–293). Psychology Press. https://doi.org/10.4324/9780203505984-16

Wilson, D. (2015). *Does Altruism Exist?* Yale University Press.

Wilson, D., Ostrom, E., & Cox, M. (2013). Generalizing the Core Design Principles for the Efficacy of Groups. *Journal of Economic Behavior & Organization*, *90S*, S21–S32. https://doi.org/10.1016/j.jebo.2012.12.010

Zhou, R., & Soman, D. (2003). Exploring the Psychology of Queuing and the Effect of the Number of People Behind. *Journal of Consumer Research*, *29*(4), 517–530. https://doi.org/10.4236/psych.2020.113033

Zipf, G. (1948). *Human Behavior and the Principle of Least Effort: An Introduction to Human Ecology*. Ravenio Books (Kindle Edition). https://doi.org/10.1002/1097-4679(195007)6:3<306::AID-JCLP2270060331>3.0.CO;2-7

6

AN ALGEBRAIC GROUP IN SOCIAL SPACE

We have gone from deriving the modified version of Weber's and Fechner's Law, to the power law, making use of both to show that the social queue and related social experiments with queue-like structure may be explained through their use, and, finally, that the social queue resource system is a social commons with a governance structure similar to that proposed by Elinor Ostrom. We finish here by showing that the social queue transform is an algebraic system satisfying the four group axioms and the additional axiom necessary to form an abelian group – the four group axioms plus the commutativity axiom of addition. Herstein (1999, p. 41) indicates that any nonempty set having an operation satisfying the four group axioms is a group, or for our purposes, a social group. Though Ring and Field axioms may be pursued by others, group axioms are sufficient to introduce the concept without wading into the full complexity of abstract algebra and further exceeding the limited capabilities of this author.

The benefit of introducing the concept of a social group as an algebraic system is that such a system is necessary if the social and mathematical sciences are to build a systematic and testable process for modeling the dynamics of complex social systems (social dynamical systems). With that said, let us return to Leon Festinger's cognitive dissonance theory to introduce one more concept before turning to show that the social queue transform satisfies the five axioms of an algebraic abelian group.

6.1 Dissonance Reduction as a Function of Time

Leon Festinger (1957, pp. 18–24) notes that when dissonance arises from a relevant social event deviating from one or more social norms, then there is a functional relationship between the strength of the pressure to reduce the dissonance and the magnitude of the dissonance. As discussed earlier, this is basically what Sherif and Sherif (1956) and Zipf (1948) determined as well. Festinger's arguments in support of his statement indicate he is not implying normalization of the deviation as a means to reduce dissonance but physical or perceptual modification of the stimuli causing the dissonance as implied in Equation 2.6 and discussed in Festinger (1957, pp. 20–21).

DOI: 10.4324/9781003325161-7

Now, consider a queue member who takes a train from New York's Grand Central station to his hometown in New Haven Connecticut every weekday at 5:00 p.m. after work. Let us ask ourselves what difference in probability of reaction might occur for this social situation if our queue member, now existing in two identical universes, experiences two simultaneous intruders in the first universe and two intruders separated by one month in the second. The argument being presented is that if the member's sensation magnitude caused by an intruder were cumulative, with no dissipation over time, then the probability of reaction by our commuting queue member toward the two intruders in the first universe, and the second intruder in the second universe, should be identical. Yet, if that were the case, seasoned Grand Central Station queue members (i.e., experiencing intrusions on a regular basis) in the (+1) position experiencing a single intruder should react with probability 1 if sensation magnitude were cumulative. We have seen empirical evidence to indicate that is not the case.

It would seem then that for a given social situation, we either reduce dissonance-causing events by appropriately reacting to the stimulus or through selective perception (interpretation). It is argued that stimuli which are significantly separated in time will result in a lower sensation magnitude than closely spaced or simultaneous stimuli. As a final thought before moving on, it is possible that normalization of deviant behavior may occur over time, assuming others in the group we identify with feel the same way. It is argued that by being exposed to repetitious low relevant events, initially considered deviant but resulting in minimal impact to the observing individual, can become normalized over time. This is a possible explanation for city dwellers and what appears to be their increased tolerance for diverse cultures and views.

6.1.1 First-Order Derivation of Dissonance Reduction as a Function of Time

In the simplest case for the relationship proposed by Festinger (1957, pp. 18–24), assume pressure to reduce the dissonance is proportional to magnitude of the dissonance with some mean proportionality constant λ. It is possible that λ is a function of the associated sensation magnitude. Other than assuming λ is constant though leads to a nonlinear differential equation that may or may not have a closed-form solution. Without known data to support otherwise, assume for now that λ is constant and a function of the social situation. Finally, assume an individual is able, over time, to reduce the stimulus causing dissonance to zero without resorting to active modification of the social environment. Based on these simplifying assumptions,

$$\frac{dI}{dt} = -\lambda \cdot I \; where \; \lambda > 0$$

since for $\lambda = 0$ change in dissonance with time is not possible

When the dissonance-causing stimulus I is introduced at time t_1, the solution to this first-order homogeneous differential equation is, *stimulus intensity $I(t) = I \cdot e^{-\lambda \cdot (t-t_1)}$ for $t \geq t_1$.*

In social situations where there is no acceptable choice for dissonance reduction, or for whatever reason dissonance cannot be fully eliminated due to a restraining force (Lewin, 1997, pp. 101, 291, 316), such as portrayed in the earlier account by Polybius of Chiomara, wife of Ortiagon, the individual will reach a certain level of dissonance and be unable to reduce the dissonance any further. To address this more complex situation, let $\lfloor I_{\infty,1} \rfloor$ be the lowest level to which an individual is able to reduce his or her dissonance over a long period of time. Assuming the dissonance was caused by a single socially deviant event in a given social situation at time t_1 having stimulus intensity $I_1(t_1)$ leads to $0 \leq \lfloor I_{\infty,1} \rfloor \leq I_1(t_1)$. Then the more general form of the equation representing stimulus intensity as a function of time may be derived using:

$$\frac{dI_1(t)}{dt} = -\lambda \cdot \left[I_1(t_1) - \lfloor I_{\infty,1} \rfloor \right].$$

The solution to this more generalized first-order differential equation is:

$$I_1(t) = \lfloor I_{\infty,1} \rfloor + \left[I_1(t_1) - \lfloor I_{\infty,1} \rfloor \right] \cdot e^{-\lambda \cdot (t-t_1)} \text{ for } t \geq t_1.$$

It is not clear how to address $\lfloor I_{\infty,n} \rfloor$ for multiple social events within the same social situation having the same type of stimuli. That must be left for future experiments or a more experienced author who has a valid argument for justifying the approach. Instead, to keep things manageable, let's settle for now on assuming $\lfloor I_{\infty,n} \rfloor = 0$ for all such events experienced by an observer.

> **Theorem 6.1**: Assume that each stimulus or simultaneous set of stimuli, all being of the same type, may be treated independently. Assume also that pressure to reduce dissonant-causing social stimuli to zero is proportional to magnitude of the dissonance with proportionality constant λ. For a given social situation and set of independent stimuli of the same type arriving at various times, with $n \in \mathbb{Z}^+$ and initial stimulus intensities represented by I_n introduced at times $t_1 \leq t_2 \leq \ldots \leq t_k \leq \ldots \leq t_n$, the addition of stimulus intensities in a queue-like system is given by:

$$I_n(t) = \left[I_n + \sum_{i=1}^{n-1} I_i(t_n - t_i) \right] \cdot e^{-\lambda \cdot (t-t_n)} \text{ for } t \geq t_n \qquad \textit{Equation 6.1}$$

Proof: This proof is performed by induction, starting with the introduction of the first stimulus or stimuli having initial stimulus intensity I_1 at time t_1, such that

$$I_1(t) = I_1 \cdot e^{-\lambda \cdot (t-t_1)} \, for \, t \geq t_1.$$

At time t_2, a second stimulus is introduced with initial intensity I_2 so that for $t \geq t_2 \geq t_1$,

$$I_2(t) = I_2 \cdot e^{-\lambda \cdot (t-t_2)} + I_1 \cdot e^{-\lambda \cdot (t-t_1)}$$

$$= \left[I_2 + I_1 \cdot e^{-\lambda \cdot (t_2-t_1)} \right] \cdot e^{-\lambda \cdot (t-t_2)} \, for \, t \geq t_2.$$

Similarly,

$$I_3(t) = I_3 \cdot e^{-\lambda \cdot (t-t_3)} + I_2 \cdot e^{-\lambda \cdot (t-t_2)} + I_1 \cdot e^{-\lambda \cdot (t-t_1)}$$

$$= \left[I_3 + I_2 \cdot e^{-\lambda \cdot (t_3-t_2)} + I_1 \cdot e^{-\lambda \cdot (t_3-t_1)} \right] \cdot e^{-\lambda \cdot (t-t_3)}$$

$$= \left[I_3 + \sum_{i=1}^{3-1} I_i \cdot e^{-\lambda \cdot (t_3-t_i)} \right] \cdot e^{-\lambda \cdot (t-t_3)} \, for \, t \geq t_3$$

Assume that this holds for stimulus intensity $I_k(t)$ which is introduced to the same individual at time t_k, then for $k \in \mathbb{Z}^+$,

$$I_k(t) = \left[I_k + \sum_{i=1}^{k-1} I_i \cdot e^{-\lambda \cdot (t_k-t_i)} \right] \cdot e^{-\lambda \cdot (t-t_k)} \, for \, t \geq t_k.$$

It remains to show that for $I_{k+1}(t), t \geq t_{k+1}$

$$I_{k+1}(t) = I_{k+1} \cdot e^{-\lambda \cdot (t-t_{k+1})} + I_k(t)$$

$$= I_{k+1} \cdot e^{-\lambda \cdot (t-t_{k+1})} + \left[I_k + \sum_{i=1}^{k-1} I_i \cdot e^{-\lambda \cdot (t_k-t_i)} \right] \cdot e^{-\lambda \cdot (t-t_k)}$$

$$= I_{k+1} \cdot e^{-\lambda \cdot (t-t_{k+1})} + \left[I_k \cdot e^{-\lambda \cdot (t_{k+1}-t_k)} + \sum_{i=1}^{k-1} I_i \cdot e^{-\lambda \cdot (t_{k+1}-t_i)} \right] \cdot e^{-\lambda \cdot (t-t_{k+1})}$$

$$= I_{k+1} \cdot e^{-\lambda \cdot (t-t_{k+1})} + \left[\sum_{i=1}^{k} I_i \cdot e^{-\lambda \cdot (t_{k+1}-t_i)} \right] \cdot e^{-\lambda \cdot (t-t_{k+1})}$$

$$= \left[I_{k+1} + \sum_{i=1}^{k} I_i \cdot e^{-\lambda \cdot (t_{k+1}-t_i)} \right] \cdot e^{-\lambda \cdot (t-t_{k+1})} \, for \, t \geq t_{k+1}$$

Therefore, by induction,

$$I_n(t) = \left[I_n + \sum_{i=1}^{n-1} I_i \cdot e^{-\lambda \cdot (t_n-t_i)} \right] \cdot e^{-\lambda \cdot (t-t_n)} \, for \, t \geq t_n \, . \qquad \blacksquare$$

To end this section, using Theorem 6.2, given the same type stimuli and social situation, the sensation magnitude for n discrete events may be stated as,

$$\sum_{i=0}^{n-1} \overline{es} \cdot ln \left[\frac{I_{i+1}(t) + \widehat{N}_0}{I_i(t) + \widehat{N}_0} \right] = \overline{es} \cdot ln \left[\frac{I_n(t) + \widehat{N}_0}{\widehat{N}_0} \right] where \, I_0(t) = 0 \, .$$

The result demonstrates that sensation magnitude under the given conditions is additive, leading to the next step of defining the algebraic social group in social space.

6.2 The Queue Transform as an Algebraic Abelian Group

The term "social space" has been bantered about since at least the late 1800s. Some credit though should be given to Pitirim Sorokin in his qualitative discussion and use of social space as mentioned in Section 1.2. What this section does is quantitatively define an algebraic group in social space K. The importance of this is that a methodical process is defined which allows for both theoretical and quantitative evaluation for confirmation or rebuttal, and, if confirmed, it creates a mathematical foundation in social space on which to expand.

Consider a specific socially deviant event that occurs in a defined social situation, with mean unit sensation magnitude $\overline{es}$ and geometric mean social noise intensity $\widehat{N}_0$ for that event. Using $I_n(t)$ as defined, Equation 6.1 leads to the next definition.

Definition 6.1: Assume that for the same social situation, social noise $\widehat{N}_0$, and stimulus type, with $I_0(t) = 0$, group member social space K, and a mean subjective dissimilarity $\Delta \overline{s}_{n,k}(t) \in K$ at time t, that

$$\left\{ \Delta \overline{s}_{n,k}(t) = \overline{s} \cdot ln \left[\frac{I_n(t) + \widehat{N}_0}{I_k(t) + \widehat{N}_0} \right] : n, k \in \mathbb{Z}^0, t \geq 0 \, with \, t \in \mathbb{R} \right\} \in K.$$

It is important to note that using $d \cong 3.76$ from Theorem 2.2 and as demonstrated in Equation 2.4, where $e \in [0,d]$,

$$\Delta\overline{s}_{n,k}(t) = \overline{s} \cdot ln\left[\frac{I_n(t) + \widehat{N}_0}{I_k(t) + \widehat{N}_0}\right] \cong \overline{es} \cdot ln\left[\frac{I_n(t) + \widehat{N}_0}{I_k(t) + \widehat{N}_0}\right] = \Delta\overline{es}_{n,k}(t) \ .$$

The latter indicates that uniform encoding and exponential encoding may be interchanged in this definition and subsequent definitions/theorems based on whether we operate in nonnegative bounded unit sensation magnitude log space or nonnegative unbounded unit sensation magnitude exponential space.

Definition 6.2: For $k,n,q \in \mathbb{Z}^0$, $\Delta\overline{s}_{n,q}, \Delta\overline{r}_{k,n}, \Delta\overline{y}_{k,q} \in K$, addition in group member social space is allowed under the following conditions,

$$\Delta\overline{s}_{n,q} + \Delta\overline{r}_{k,n} = \Delta\overline{y}_{k,q}$$

$$\Delta\overline{s}_{n,k} + \Delta\overline{r}_{k,q} = \Delta\overline{y}_{n,q} \ .$$

Theorem 6.2: Assume the same social situation and stimulus type. Also assume, without loss of generality, that $\widehat{N}_0$ is constant for each subsequent intrusion. Let $k,m,n,q \in \mathbb{Z}^0$ and $\Delta\overline{s}_{n,q}, \Delta\overline{r}_{k,n}, \Delta\overline{y}_{k,q} \in K$, then the queue transform, supported by field axioms for addition and multiplication in the real number system, meets the five algebraic group axioms (GA) of an abelian group as defined by Herstein (1999, pp. 40–43) under the operation of logarithmic addition. Using Definitions 6.1 and 6.2,

GA 1) (Closed under Addition): If $\Delta\overline{r}_{k,n}, \Delta\overline{s}_{n,q}, \in K$ then $\Delta\overline{r}_{k,n} + \Delta\overline{s}_{n,q} \in K$. Similarly,

if $\Delta\overline{r}_{k,q}, \Delta\overline{s}_{n,k}, \in K$ then $\Delta\overline{r}_{k,q} + \Delta\overline{s}_{n,k} \in K$.

GA 2) (Commutative): $\Delta\overline{r}_{k,n} + \Delta\overline{s}_{n,q}, = \Delta\overline{s}_{n,q} + \Delta\overline{r}_{k,n}$.

GA 3) (Associative): $(\Delta\overline{r}_{k,n} + \Delta\overline{s}_{n,q}) + \Delta\overline{y}_{q,m} = \Delta\overline{r}_{k,n} + (\Delta\overline{s}_{n,q} + \Delta\overline{y}_{q,m})$.

GA 4) (Identity): There exists an $e \in K$ such that $\Delta\overline{s}_{n,k} + e = e + \Delta\overline{s}_{n,k} = \Delta\overline{s}_{n,k}$ *for all* $\Delta\overline{s}_{n,k} \in K$

GA 5) (Inverse): For every $\Delta\overline{s}_{n,k} \in K$, there exists $\Delta\overline{r}_{q,n} \in k$ such that $\Delta\overline{s}_{n,k} + \Delta\overline{r}_{q,n} = \Delta\overline{r}_{q,n} + \Delta\overline{s}_{n,k} = e$.

Proof: By direct proof,
GA 1) To prove $\Delta\overline{r}_{k,n} + \Delta\overline{s}_{n,q} \in K$. Proving $\Delta\overline{r}_{k,q} + \Delta\overline{s}_{n,k} \in K$ may be shown in a similar manner and is therefore left to the reader.

$$\Delta \bar{r}_{k,n} + \Delta \bar{s}_{n,q} = \bar{s} \cdot ln \left[\frac{I_k(t) + \widehat{N}_0}{I_n(t) + \widehat{N}_0} \right] + \bar{s} \cdot ln \left[\frac{I_n(t) + \widehat{N}_0}{I_q(t) + \widehat{N}_0} \right]$$

$$= \bar{s} \cdot \left(ln \left[\frac{I_k(t) + \widehat{N}_0}{I_n(t) + \widehat{N}_0} \right] + ln \left[\frac{I_n(t) + \widehat{N}_0}{I_q(t) + \widehat{N}_0} \right] \right) \cdot$$

$$= \bar{s} \cdot ln \left[\frac{I_k(t) + \widehat{N}_0}{I_q(t) + \widehat{N}_0} \right]$$

But $\bar{s} \cdot ln \left[\dfrac{I_k(t) + \widehat{N}_0}{I_q(t) + \widehat{N}_0} \right] \in K$ by definition; hence, $\Delta \bar{r}_{k,n} + \Delta \bar{s}_{n,q} \in K$.

GA 2) To prove $\Delta \bar{r}_{k,n} + \Delta \bar{s}_{n,q} = \Delta \bar{s}_{n,q} + \Delta \bar{r}_{k,n}$:

$$\Delta \bar{r}_{k,n} + \Delta \bar{s}_{n,q} = \bar{s} \cdot ln \left[\frac{I_k(t) + \widehat{N}_0}{I_n(t) + \widehat{N}_0} \right] + \bar{s} \cdot ln \left[\frac{I_n(t) + \widehat{N}_0}{I_q(t) + \widehat{N}_0} \right]$$

$$= \bar{s} \cdot ln \left[\frac{I_k(t) + \widehat{N}_0}{I_q(t) + \widehat{N}_0} \right], and$$

$$\Delta \bar{s}_{n,q} + \Delta \bar{r}_{k,n} = \bar{s} \cdot ln \left[\frac{I_n(t) + \widehat{N}_0}{I_q(t) + \widehat{N}_0} \right] + \bar{s} \cdot ln \left[\frac{I_k(t) + \widehat{N}_0}{I_n(t) + \widehat{N}_0} \right] \cdot$$

$$= \bar{s} \cdot ln \left[\frac{I_k(t) + \widehat{N}_0}{I_q(t) + \widehat{N}_0} \right]$$

Hence, $\Delta \bar{r}_{k,n} + \Delta \bar{s}_{n,q} = \Delta \bar{s}_{n,q} + \Delta \bar{r}_{k,n}$.

GA 3) To prove $(\Delta \bar{r}_{k,n} + \Delta \bar{s}_{n,q}) + \Delta \bar{y}_{q,m} = \Delta \bar{r}_{k,n} + (\Delta \bar{s}_{n,q} + \Delta \bar{y}_{q,m})$:

$$(\Delta \bar{r}_{k,n} + \Delta \bar{s}_{n,q}) = \bar{s} \cdot ln \left[\frac{I_k(t) + \widehat{N}_0}{I_q(t) + \widehat{N}_0} \right], so$$

$$(\Delta \bar{r}_{k,n} + \Delta \bar{s}_{n,q}) + \Delta \bar{y}_{q,m} = \bar{s} \cdot ln \left[\frac{I_k(t) + \widehat{N}_0}{I_q(t) + \widehat{N}_0} \right] + \bar{s} \cdot ln \left[\frac{I_q(t) + \widehat{N}_0}{I_m(t) + \widehat{N}_0} \right]$$

$$= \bar{s} \cdot ln \left[\frac{I_k(t) + \widehat{N}_0}{I_m(t) + \widehat{N}_0} \right]$$

Similarly,

$$\Delta\bar{r}_{k,n} + (\Delta\bar{s}_{n,q} + \Delta\bar{y}_{q,m}) = \bar{s} \cdot ln\left[\frac{I_k(t) + \widehat{N}_0}{I_n(t) + \widehat{N}_0}\right] + \left(\bar{s} \cdot ln\left[\frac{I_n(t) + \widehat{N}_0}{I_q(t) + \widehat{N}_0}\right]\right.$$

$$\left. + \bar{s} \cdot ln\left[\frac{I_q(t) + \widehat{N}_0}{I_m(t) + \widehat{N}_0}\right]\right)$$

$$= \bar{s} \cdot ln\left[\frac{I_k(t) + \widehat{N}_0}{I_n(t) + \widehat{N}_0}\right] + \bar{s} \cdot ln\left[\frac{I_n(t) + \widehat{N}_0}{I_m(t) + \widehat{N}_0}\right]$$

$$= \bar{s} \cdot ln\left[\frac{I_k(t) + \widehat{N}_0}{I_m(t) + \widehat{N}_0}\right].$$

Hence

$$(\Delta\bar{r}_{k,n} + \Delta\bar{s}_{n,q}) + \Delta\bar{y}_{q,m} = \bar{s} \cdot ln\left[\frac{I_k(t) + \widehat{N}_0}{I_m(t) + \widehat{N}_0}\right]$$

$$= \Delta\bar{r}_{k,n} + (\Delta\bar{s}_{n,q} + \Delta\bar{y}_{q,m}).$$

GA 4) To prove there exists an $e \in K$ such that

$$\Delta\bar{s}_{n,k} + e = e + \Delta\bar{s}_{n,k} = \Delta\bar{s}_{n,k} \text{ for all } \Delta\bar{s}_{n,k} \in K.$$

Using the same approach as in GA 1), we must find a value $q \in \mathbb{Z}^0$ such that

$$\Delta\bar{r}_{k,n} + \Delta\bar{s}_{n,q} = \Delta\bar{r}_{k,n} + \bar{s} \cdot ln\left[\frac{I_n(t) + \widehat{N}_0}{I_q(t) + \widehat{N}_0}\right] = \Delta\bar{r}_{k,n}.$$

Based on Definitions 6.1 and 6.2, q must equal n. Therefore, relying on the field axioms for multiplication,

$$\Delta\bar{s}_{n,k} + e = e + \Delta\bar{s}_{n,k} = \Delta\bar{s}_{n,k} \text{ for all } \Delta\bar{s}_{n,k} \in K.$$

GA 5) To prove for every $\Delta\bar{s}_{n,k} \in K$, there exists an $\Delta\bar{r}_{q,n} \in K$ such that $\Delta\bar{s}_{n,k} + \Delta\bar{r}_{q,n} = \Delta\bar{r}_{q,n} + \Delta\bar{s}_{n,k} = e$, where we have already shown that $e = 0$.

$$\Delta \bar{s}_{n,k} + \Delta \bar{r}_{q,n} = \bar{s} \cdot ln \left[\frac{I_n(t) + \hat{N}_0}{I_k(t) + \hat{N}_0} \right] + \bar{s} \cdot ln \left[\frac{I_q(t) + \hat{N}_0}{I_n(t) + \hat{N}_0} \right]$$

$$= \bar{s} \cdot ln \left[\frac{I_q(t) + \hat{N}_0}{I_k(t) + \hat{N}_0} \right] but$$

$$\bar{s} \cdot ln \left[\frac{I_q(t) + \hat{N}_0}{I_k(t) + \hat{N}_0} \right] = 0 \; if \; and \; only \; if$$

$$q = k \; so \; that \; \Delta \bar{s}_{n,k} + \Delta \bar{r}_{k,n} = 0 = e.$$

Hence, having already proven commutativity, for every $\Delta \bar{s}_{n,k} \in K$, there exists an $\Delta \bar{r}_{q,n} \in K$ such that $\Delta \bar{s}_{n,k} + \Delta \bar{r}_{q,n} = \Delta \bar{r}_{q,n} + \Delta \bar{s}_{n,k} = e.$ ∎

The remainder of this section will discuss the implications and examples of each item proven in Theorem 6.2 in the context of a queue. With that, consider a queue having ten members, where the member position of interest is third from the front. As a result, there are two members ahead of the member of interest and seven members behind the member of interest. Until an intruder arrives, the queue is operating as members believe it should, so $n = 0$ and the stimulus intensity $I_n = I_0 = 0$. Based on what commodity the members are waiting for in queue, and based on the social situation, the mean queue unit sensation magnitude is $\bar{s}$. Mean noise for the particular social situation is $\hat{N}_0$. As before, it is assumed the subject member of interest does not react in an attempt to modify the social environment.

<u>Group Axiom 1 Example</u>: At time t_1, an intruder arrives and steps in front of the member of interest causing initial stimulus intensity I_1. At $t_2 = 5$ minutes, two additional simultaneous intruders join the queue between the member of interest and the first intruder resulting in a cumulative stimulus intensity at time $t \geq t_2$ of $I_2(t)$. The initial stimulus intensity of the two simultaneous intruders has intensity I_2. Then, as per Theorem 6.1,

$$I_2(t \geq 5 \; min) = \left[I_2 + I_1(5 \; min) \right] \cdot e^{-\lambda \cdot (t - 5 \; min)}$$

The sensation magnitude at time $t_2 = 5 \; min$ of the second intrusion is:

$$\bar{s} \cdot ln \left[\frac{I_1(5 \; min) + \hat{N}_0}{\hat{N}_0} \right] \cdot ln \left(e \cdot \frac{7 + N_{UA}}{N_{UA}} \right) + \bar{s} \cdot ln \left[\frac{I_2(5 \; min) + \hat{N}_0}{I_1(5 \; min) + \hat{N}_0} \right]$$

$$\cdot ln \left(e \cdot \frac{7 + N_{UA}}{N_{UA}} \right) = \bar{s} \cdot ln \left[\frac{I_2(5 \; min) + \hat{N}_0}{\hat{N}_0} \right] \cdot ln \left(e \cdot \frac{7 + N_{UA}}{N_{UA}} \right).$$

For sensation magnitude at time $t \geq t_2$,

$$\overline{s} \cdot ln\left[\frac{I_2(t \geq 5 \ min) + \widehat{N}_0}{\widehat{N}_0}\right] \cdot ln\left(e \cdot \frac{7 + N_{UA}}{N_{UA}}\right)$$

$$\leq \overline{s} \cdot ln\left[\frac{I_2(5 \ min) + \widehat{N}_0}{\widehat{N}_0}\right] \cdot ln\left(e \cdot \frac{7 + N_{UA}}{N_{UA}}\right).$$

Group Axion 2 will be skipped since it is trivial.

Group Axiom 3 Example: Building on the Group Axiom 1 example, a third single intruder now arrives at time $t_3 = 15 \ min$ and steps into the queue between the member of interest and the first three intruders. The first two intruder events were calculated to have a cumulative stimulus intensity at time t of $I_2(t \geq 5 \ min)$. Now, with the third event resulting in a fourth intruder into queue, the cumulative stimulus intensity at time $t \geq 15 \ min$ as per Theorem 6.1 is

$$I_3(t \geq 15 \ min) = I_3(t) = \left[I_3 + \sum_{i=1}^{3-1} I_i \cdot e^{-\lambda \cdot (t_3 - t_i)}\right] \cdot e^{-\lambda \cdot (t - t_3)}.$$

The sensation magnitude now felt by the member of interest at time $t_3 = 15 \ min$ is

$$\overline{s} \cdot ln\left[\frac{I_2(15 \ min) + \widehat{N}_0}{\widehat{N}_0}\right] \cdot ln\left(e \cdot \frac{7 + N_{UA}}{N_{UA}}\right)$$

$$+ \overline{s} \cdot ln\left[\frac{I_3(15 \ min) + \widehat{N}_0}{I_2(15 \ min) + \widehat{N}_0}\right] \cdot ln\left(e \cdot \frac{7 + N_{UA}}{N_{UA}}\right)$$

$$= \overline{s} \cdot ln\left[\frac{I_3(15 \ min) + \widehat{N}_0}{\widehat{N}_0}\right] \cdot ln\left(e \cdot \frac{7 + N_{UA}}{N_{UA}}\right).$$

With sensation magnitude at time $t \geq t_3 = 15 \ min$ we have,

$$\overline{s} \cdot ln\left[\frac{I_3(t \geq 15 \ min) + \widehat{N}_0}{\widehat{N}_0}\right] \cdot ln\left(e \cdot \frac{7 + N_{UA}}{N_{UA}}\right)$$

$$\leq \overline{s} \cdot ln\left[\frac{I_3(15 \ min) + \widehat{N}_0}{\widehat{N}_0}\right] \cdot ln\left(e \cdot \frac{7 + N_{UA}}{N_{UA}}\right).$$

Group Axiom 4 Example: If at time $t_{n+1} > t_n$, the queue member of interest experiences no additional stimulus intensity, the equivalent expression would be,

$$\bar{s} \cdot ln\left[\frac{I_n(t_{n+1}) + \hat{N}_0}{\hat{N}_0}\right] \cdot ln\left(e \cdot \frac{7 + N_{UA}}{N_{UA}}\right)$$

$$+ \bar{s} \cdot ln\left[\frac{I_n(t_{n+1}) + \hat{N}_0}{I_n(t_{n+1}) + \hat{N}_0}\right] \cdot ln\left(e \cdot \frac{7 + N_{UA}}{N_{UA}}\right)$$

$$= \bar{s} \cdot ln\left[\frac{I_n(t_{n+1}) + \hat{N}_0}{\hat{N}_0}\right] \cdot ln\left(e \cdot \frac{7 + N_{UA}}{N_{UA}}\right).$$

Interpretation: In this case, no additional intruders have entered the queue, and none have left the queue. In effect, nothing has changed between time t_n and t_{n+1} except the reduction in dissonance brought about by the increment of time $t_{n+1} - t_n$.

Group Axiom 5 Example: This final axiom leads into a subtraction of stimulus intensities (intruders) from the social situation. For this example, assume that a total of n intruder groups (i.e., one or more simultaneous intruders) have entered the queue at various times. If at time t_{n+1} all of the intruders turn around, apologize to the member of interest, and then leave, we will assume for this extreme case that dissonance previously experienced by the member of interest due to the intruders is reduced to zero. We then have,

$$\bar{s} \cdot ln\left[\frac{I_n(t_{n+1}) + \hat{N}_0}{\hat{N}_0}\right] \cdot ln\left(e \cdot \frac{7 + N_{UA}}{N_{UA}}\right)$$

$$+ \bar{s} \cdot ln\left[\frac{\hat{N}_0}{I_n(t_{n+1}) + \hat{N}_0}\right] \cdot ln\left(e \cdot \frac{7 + N_{UA}}{N_{UA}}\right) =$$

$$\bar{s} \cdot ln\left[\frac{I_n(t_{n+1}) + \hat{N}_0}{\hat{N}_0}\right] \cdot ln\left(e \cdot \frac{7 + N_{UA}}{N_{UA}}\right) - \bar{s}$$

$$\cdot ln\left[\frac{I_n(t_{n+1}) + \hat{N}_0}{\hat{N}_0}\right] \cdot ln\left(e \cdot \frac{7 + N_{UA}}{N_{UA}}\right) = 0.$$

Interpretation: What dissonance might exist after the intruders leave would not be from their breaking a social norm, but from trying to interpret what just happened and why. So, one source of dissonance removed might result in another form of dissonance, which in this case would likely be much smaller in magnitude.

6.3 Potential Implications of Ambiguity Reduction, Cumulative Dissonance, and/or Reevaluation of the Social Situation

In Section 6.1 it was assumed that $\widehat{N}_0$ remains constant as each subsequent intruder or intruders enter the queue ahead of the member of interest. In reality, as intruders continue to enter ahead of the member of interest within the waiting time of the queue, it will become readily apparent to the queue member that he or she is being taken advantage of in a methodical manner. With this individual reduction in social noise (i.e., reduction of ambiguity and/or uncertainty), the cumulative effect, and/or possibly a reevaluation of the social deviation observed, the sense of moral outrage resulting from this clear violation of a social norm would increase based on our equation for sensation magnitude and the role that $\widehat{N}_0$ plays. Oberholzer-Gee (2006, pp. 438) conducted an ad hoc field experiment in 2002 that demonstrates this concept quite dramatically. Whether the increase in probability of reaction he experienced was due to decreased noise based on reduced ambiguity, increased sensation magnitude due to reinterpretation of the social deviation, the cumulative effect of stimuli, or some combination of all three could not be ascertained from the data. Future design of experiments allowing for the necessary data would be of interest in quantitatively validating or disproving this last hypothesis and theorems within this chapter.

References

Festinger, L. (1957). *A Theory of Cognitive Dissonance*. Stanford University Press. Retrieved from www.sup.org/books/title/?id=3850

Herstein, I.N. (1999). *Abstract Algebra*. John Wiley & Sons. Retrieved from www.wiley.com/en-us/Abstract+Algebra%2C+3rd+Edition-p-9780471368793

Lewin, K. (1997). In G.W. Lewin (Ed.), *Resolving Social Conflicts and Field Theory in Social Science*. American Psychological Association (Kindle Edition). Retrieved from www.apa.org/pubs/books/4318600

Oberholzer-Gee, F. (2006). A Market for Time Fairness and Efficiency in Waiting Lines. *KYKLOS, 59*(3), 427–440. https://doi.org/10.1111/j.1467-6435.2006.00340.x

Sherif, M., & Sherif, C. (1956). *An Outline of Social Psychology*. Harper & Brothers.

Zipf, G. (1948). *Human Behavior and the Principle of Least Effort: An Introduction to Human Ecology*. Ravenio Books (Kindle Edition). https://doi.org/10.1002/1097-4679(195007)6:3<306::AID-JCLP2270060331>3.0.CO;2-7

7

HISTORY AS DATA

Without a fair, orderly, and efficient method of allocating resource units, local appropriators have little motivation to contribute to the continued provision of the resource system.
Dr. ElinorOstrom, 1990, p. 33, with permission Cambridge University Press

A significant amount of material has been covered in the previous six chapters. Much of it is grounded in established theory and supported by empirical data. More importantly, it can either be proven or disproven, and, if disproven, improved upon or discarded as more information is gained. This is possible through the quantitative and testable structure now available for investigating certain aspects of social psychology in a more systematic quantitative manner. Additionally, through the use of the proposed algebraic social space, theoretical manipulation of our social environment may be accomplished for certain social situations and then tested in field experiments or eventually evaluated against historical data. It is not envisioned that everything as presented here is fully mature, but that through further experimental results and academically diverse views, the maturity of concepts and equations may be improved. In essence, it is believed that this work allows for a more systematic approach toward evaluating certain aspects of social psychological theories and their possible interrelationships.

This final chapter considers the perennial issue of social status involving dominant groups and subordinate groups (Tajfel and Turner, 2001, p. 98). As proposed by Ostrom (1990), status should be the reward for significant contributions to the community and should therefore be conferred by the community based on the risk taken and resultant benefit provided to the community. There are two extremes though that seemingly elude a social solution. The first involves those that do not or cannot contribute to the community and are in essence free-riders by either choice or fate. The second involves those that accumulate wealth beyond any reasonable measure in relation to the relative wealth of the community, yet the community and common resource units are needed to create that particular wealth. Without putting wealth back into the

DOI: 10.4324/9781003325161-8

community in the form of social benefit and supporting infrastructure that allows the wealth, then the wealthy in effect become free-riders. Too much of either extreme leads to the eventual collapse of the resource system. How to control the accumulation of wealth and consumption of resource units while still adequately rewarding those who take greater risks or expend greater effort for the benefit of the community is the key question. In terms of governance, Design Principles 2 (Appropriation and provision) and 4 (Monitoring the resource) appear to be the first-order elements for understanding and investigating the implications and eventually finding a solution to controlling the effect of free-riding and its negative impact on the stability of a social commons. Our focus in this chapter is directed at a defined community with those in high status who do not provide adequate provision for the community resource(s) they consume and enjoy. To pursue this concept, viewing history as data, let's consider social events leading to the revolution of 1848 within the German Confederation.

7.1 The German Confederation and Social Events Leading to Its 1848 Revolution

There seems to be differing opinions as to what led to the 1848 revolution within the German Confederation, with a good summary and interesting economic analysis provided by Berger and Spoerer (2001, p. 295) as to what they believe the deciding event was. When trying to understand historical events, or current events from a historical perspective, it is rare that a single explanation exists as to why something unfolded with reaction ultimately occurring the way it did. History leading up to an event provides the context of any social situation. Similarly, community members within the developing social situation, when applicable, use associated past history as passed to them to recognize and help interpret current social events and appropriate reactions to those events. We have observed over the past few chapters that when one or more beliefs regarding what should be differs from what is, dissonance occurs, and as the dissonance increases, the pressure to react and reduce that dissonance also increases. Therefore, instead of looking for a single cause leading to a reaction, we need to understand what led to increasing community (i.e., German Confederation) pressure for change by focusing on the development of the social situation within the community of interest, understanding the evolving social norms, and finally consider through firsthand accounts what event or events facilitated transforming the social pressure into social reaction or, in this case, revolution. Once this is complete, then the application of the queue transform will be considered as a possible but currently limited quantitative means of explanation.

It will be shown, using historical data to develop a sense of the social situation promoting the German Confederation's revolution of 1848, that the primary event converting existing social pressure into revolution as witnessed by those who were there at the time was the February 1848 French Revolution. Supporting this thesis requires a historical backdrop of associated events starting with the Napoleonic Wars ending in 1814, insurrections in the German Confederation immediately after the 1830 Paris Revolution, external competition and the growth of internal industries requiring the need for unification of the Confederation states, the general Evangelical Synod of

1846 (Jensen, 1974), the United Diet of 1847 using letters from the Earl of Westmo-reland, combined with Carl Schurz's autobiography (Schurz, 1907), and the bad crop years and food riots of 1845 through 1847 (Gailus, 1994). With these backdrops in place, evaluation and interpretation of emigration data from 1825 through 1849 are performed as an additional means to understand how and why the revolution occurred when it did (Pfaff and Kim, 2003). To effectively make the argument, a more compre-hensive definition for revolution must first be developed, a definition that goes beyond the traditional Merriam-Webster definition. To this end, a definition for our purposes will be developed based on the more thorough primary source observations of Karl Marx (Marx, 1851, October 25) and John Maynard Keynes (Keynes, 1919).

7.1.1 Defining Revolution as a Tragedy of the Commons

Karl Marx famously provided a post analysis of the 1848 French Revolution and its aftermath in 1852 (Marx, 1852). He similarly provided a more detailed post analysis of the German Confederation's 1848–1849 revolution – published through multiple articles from 1851 to 1852 for the New-York Daily Tribune while in London – as part of a series covering events in Europe which were later combined (Marx, 1912). In his 25 October 1851 New-York Daily Tribune article addressing the German revolution of 1848, Marx states, "Everyone knows nowadays that whenever there is a revolution-ary convulsion, there must be some social want in the background, which is prevented by outworn institutions from satisfying itself" (Marx, 1851, October 25, p. 6). He goes on to state in this same article,

> That the sudden movements of February and March 1848 were not the work of single individuals, but spontaneous, irresistible manifestations of national wants and necessities, more or less clearly understood, but very distinctly felt by numerous classes in every country, is a fact recognized everywhere.
>
> *(Marx, 1851, October 25, p. 6)*

It is critical to note that Marx was focused on the national perspective of wants and needs, which is equivalent to the community of interest and its resource systems, and not on any individual wants and needs. This distinction will be emphasized when we consider emigration data.

Sixty-eight years later, after the First World War, John Maynard Keynes provided additional insight into what supports a revolution and therefore how it may be fur-ther defined. Keynes, an official representative of the British Treasury at the Paris Peace Conference up until 7 June 1919, resigned his position in protest against the Treaty of Peace and the conditions it imposed upon the defeated German Republic. In his prescient book written that same year, presenting logical arguments against the terms of the Treaty of Peace, Keynes supplemented Marx's statement on the cause of revolution by noting a belief at the Paris Peace Conference that, "The only safeguard against Revolution in Central Europe is indeed the fact that, even to the minds of men who are desperate, Revolution offers no prospect of improvement whatever" (Keynes, 1919, loc. 2818). The reference is to the fact that both Germany and Russia were being blocked from economic recovery by the Allies after the war, the Allies believing this approach would reduce the risk of revolution and a subsequent union between Germany and Russia (Keynes, 1919, loc. 2795). What was imposed on Germany therefore was not an internal injustice that might be remedied through internal revolution but an external injustice that as Keynes foresaw, "will affect everyone in the long-run, but perhaps not in the way that is striking or immediate" (Keynes, 1919, loc. 2821). The Treaty of Peace in effect was an externally imposed feudal system, and Keynes understood the implications this could have for Germany.

Taking these two perspectives into account, revolution will be defined as a national level social reaction caused by sufficient social pressure resulting from unmet national level wants and needs. It must be a social reaction which offers sufficient prospect for improvement against political institutions (i.e., Axiom 2) that otherwise deny the social wants and needs of its citizens. To show there is no contradiction for the extreme case, if survival is at stake for those already revolting, the revolution will exist based on the belief that it is the only option which offers a chance for improvement to those already involved.

7.1.2 Establishing and Legitimizing Representation as a Group Social Norm

Though applicable in general, for states within the German Confederation, the basis of unrest during the 1830 to 1833 insurrections and the more famous revolution of 1848–1849 evolved from the unmet social wants and needs of its citizens. The driving force behind the insurrections and revolution is argued to be the belief that unification of the German states under a representative constitution would improve social and economic conditions. A representative constitution had been promised to Prussians in 1813 by King Frederick William III in preparation for the final set of military campaigns that led to defeating Napoleon (Marx, 1851, October 28; Schurz, 1907). After Napoleon's defeat in 1814, a German Confederation consisting of 38 independent states was established, controlled by nobility, and loosely tied together under the Federal Act of 1815 supporting protection of the Confederation and its nobility (Deutsche Bundesakte, 1815). The promise of a representative government was still just a promise, as was that of an 1815 promise for a free press, both of which were later incorporated into law in 1820 which were either subsequently revoked or implemented in a meaningless manner (Marx, 1851, October 28). This desire for a representative government continued after 1820 by those advocating for unification and representation, as did the desire to maintain the existing Confederation by many within the Prussian civil service, many of whom were aristocrats who benefitted from the existing political institution and the resources it provided to them.

The seed of this dichotomy began to clearly emerge after Prussia's defeat by Napoleon in 1806, thus bringing the Holy Roman Empire to an end. Being defeated by Napoleon, Prussia, through many of its Ministers, recognized that changes to the social structure were needed – in particular, the feudal system had become bankrupt, nobility had become free-riders, and the institution of serfdom needed to be eliminated in order for peasants to cultivate their own land and develop a sense of national identity with the fatherland. It was believed this in turn could be used to build an army of peasants willing to fight for their land and their country as envisioned by academics such as Ernst Moritz Arndt, a well-traveled history professor who originated from peasantry, those in the military such as then middle-rank officer Carl von Clausewitz and his mentor Prussian General Gerhard Johann David von Scharnhorst (Gagliardo, 1969, Chp. 8), and some in the Prussian Ministry such as the lawyer and anonymous publisher Friedrich von Coelln, who was highly critical of the free-riding nobility and who noted among many other injustices:

> At the beginning of the feudal system, the noble knight who paid no taxes was inexpensive since he fought for the state with great valor, while the serf's war services were not nearly as significant. Now, however, the serf is forced to go to war on his own and also has to pay land taxes. This is not right.
>
> *(Coelln, 1807, p. 31)*

Through defeat by Napoleon, humiliation, and the resulting national dissonance, sufficient pressure for social change to address an identified need had finally been achieved.

To begin addressing necessary change, King Frederick William III signed The Prussian Reform Edict on 9 October 1807 which introduced land ownership reform and abolished serfdom in Prussia. This was quickly replicated in the remaining states. The Edict, though in more words, basically states the ruling nobility has no other choice than to grant individual freedom to its serfs if it wants to survive (William III, 1807/1902, pp. 27–30). The Prussian Reform Edict and the ensuing social change led to the envisioned advantage during the Wars of Liberation against Napoleon in 1813 and 1814. The edict, though successful in its intended purpose, also led to land speculation and an unintended crash of the land market in 1821 due to foreign improvements in agricultural production which lowered prices (Burgdorfer, 1931, p. 342), lack of credit for improvements or purchasing of land (Frederiksen, 1894), and peasants who were unprepared or unable to compete having to leave their land. This was the beginning of the Agrarian Crisis, a time of uncertainty for many (Gagliardo, 1969, Chp. 8), and after food prices began to increase, a precursor to insurrections from 1830–1833 which may be viewed as the source of collective memory and a trial run for the revolution of 1848–1849.

It is shortly after the Paris Revolution of July 1830 that planned large-scale insurrections within the German Confederation began. The first public disturbance reported by the British envoy George William Chad in Frankfurt to the British Foreign Secretary, George Hamilton-Gordon, the 4th Earl of Aberdeen, involves a riot in Hesse-Darmstadt that immediately disbursed on the arrival of soldiers. George Chad attributes the riot to the increasing price of bread resulting from anticipation of war (Chad, 1830, Frankfurt, September 9). In another 11 September 1830 letter from Frankfurt by George Chad to the 4th Earl of Aberdeen, a possible insurrection in Brunswick is noted, and on 22 September 1830 a disturbance in Carlsruhe, in Wurttemberg near the border with Baden, occurs due to Wurttemberg signing a commercial treaty with Prussia (Chad, 1830, Frankfurt, September 11 and 22; Bavarian thaler,

1829). In the most telling of the string of letters, the 29 September 1830 letter from the British envoy John Milbanke in Frankfurt to the Earl of Aberdeen describes how close the insurrections, in parts of Hesse, were to becoming a revolution within the state of Hesse. In fact, the Hesse government began creating a constitution modeled after that of France's (Milbanke, 1830, Frankfurt, September 29). Finally, Milbanke notes that between events in Brussels and uprisings in the German states, the opportunity for revolution within the German Confederation was real (Milbanke, 1830, Frankfurt, September 29).

Unrest in many of the German Confederation States continued seemingly uncoordinated throughout 1831. Rioting by the lower class was attributed to economic grievances that included taxation and the price of bread (needs), but leadership (uncertainty information) was provided by the upper middle class, particularly in the southwestern states (Sperber, 1989). The British envoy Thomas Cartwright sent a report on 16 January 1831 concerning an insurrection in Gottingen directed at obtaining a representative constitution. On 17 February 1831, he noted in his letter that King Ludwig of Bavaria issued an edict to regulate the press (contrary to Design Principle 4 and the need for monitoring) in an effort to prevent further disturbances in that state, and after mentioning additional disturbances and potential plots, he ends the year with his letter concerning Prussia and Austria's combined efforts to severely limit the press which they view as a dangerous weapon wielded by the Liberal Party (Cartwright, 1831, Frankfurt, November 23). The disturbances and unrest continue into 1832 and, from the Diet's viewpoint, reach a peak the middle of that year. In May of 1832, a large festival, the largest of many smaller such festivals held in the southern Rhineland over the previous 2 years, occurred around Hambach Castle in Neustadt an der Weinstraße, which is just west of the Rhine River in what was then part of Bavaria and northwest of Stuttgart. Termed the Hambach Festival, it attracted thousands of people over the four-day period 27 to 30 May. This quasi-revolutionary gathering was brought about through a longer-term opposition political campaign, unjust taxation, tariff policies, exacerbated by a series of poor harvests. Johann August Wirth, a radical democratic journalist, spoke out during the festival, stating much as Friedrich von Coelln had 20 years earlier that the German Confederation's 34 states (i.e., principalities) were being exploited by nobility that who were not providing benefit to their community (Wirth, 1832). He went on to lambast the absolutist alliance of Prussia, Austria, and Russia and decried how the three were responsible for maintaining their monarchies while preventing surrounding countries from gaining their liberty and any reasonable form of governmental institutions. What was most interesting was his foretelling of how Germany may one day become a democratic government through the people according to their needs (Wirth, 1832). Shortly after the Hambach Festival, and partially in response, the Confederal General Assembly of 1832 promulgated the Six Articles of 28 June 1832 and then the Ten Articles of 5 July 1832 (Six and Ten Articles, 1832).

The Six and Ten Articles were a reaction by nobility to the Paris Revolution of 1830, subsequent riots and insurrections in the Confederation, and the Hambach Festival. In the Six Articles, the Confederal General Assembly more firmly enforced allegiance of the Confederation states to the Confederation and its Constitution. The Ten Articles were instead directed at repressing opportunity for future citizen-riots within the

Confederation. The most noteworthy of these ten articles may be summarized as: 1) government approval was now required for distribution of non-German Confederation newspapers less than 20 pages [meant to stop distribution of foreign pamphlets]; 2) all political associations or non-customary popular assemblies and festivals are prohibited; and 3) military assistance from surrounding states may be requested against large popular movements. It is Article 6 of the Ten Articles (Six and Ten Articles, 1832) that creates a police state atmosphere, enforcing repression and collective memory as discussed in Pennebaker and Banasik (1997), a collective memory and strengthening of group cohesion that is put to use by advocates for a representative government over the following 16 years.

It is Articles 4 and 6 of the Six Articles that were considered by the British envoy in Frankfurt to present the greatest threat to Confederation state sovereignty as originally provided in the Federal Act of 1815 under Article 2 (Deutsche Bundesakte, 8 Juni 1815). In the Six Articles, Article 4 removes protection for the independence and inviolability of individual German states (Six and Ten Articles, 1832), of which the southwestern German States were most likely to react against removal of such a protection, given their historical independence from Prussia under Napoleon. Article 6 of the Six Articles (Six and Ten Articles, 1832) completes this coup de grace by stating legal interpretation of the Confederal and Final Act would be up to the German Confederation through the Federal Assembly, which was basically under the control of Prussia and Austria.

Thomas Cartwright wrote two lengthy letters analyzing the Six Articles for Viscount Palmerton. He was concerned that the wording may be construed by the independent states as an attempt by Prussia and Austria to gain greater control over their countries, control far exceeding what was agreed to in the original Federal Act of 1815. From his analysis, he made two very astute observations. The first, in his 25 June letter, was that the southwestern states of the Confederation might revolt (Cartwright, 1832, Frankfurt, June 25). The second, in a letter written on 16 July 1832, adds to the first by stating that some of the more liberal states, particularly the southwestern Confederation states, might turn to France if their state liberties are significantly threatened (Cartwright, 1832, Frankfurt, July 16).

In effect, the German Confederation through Prussia and Austria was becoming overly repressive against other states, to include the southwestern states of Baden, Wurttemberg, and Bavaria, that otherwise desired independence and a certain level of individual freedom as was introduced during the French occupation under Napoleon between 1806 and 1813 and which they were loath to relinquish. Alternative concepts for governance in the Confederation were obviously in existence as evidenced through certain insurrections (Cartwright, 1833, Frankfurt, April 5), with the government actively trying to control their spread. But alternatives to the Confederation, while being repressed in speech, were also being pursued in commerce. By the mid-1820s, custom unions (i.e., Zollverein) were being established between certain states to promote trade, an indirect approach toward German political unity and a process which had certain economic benefit for the slowly expanding industrial merchants within the German Confederation States (Dusseldorfer Zeitung, 1843). After the Paris Revolution in 1830 and resultant insurrections within the German

Confederation, further repressive measures led to fewer liberties for the citizens and the press, a direction counter to King Frederick William III's past promises of a representative constitution and a free press. Yet, the emerging industrial sector needed unification to further grow and compete, and unification required a national level government. The belief by commoners and industry leaders alike, of what should be and what they were observing, was steadily diverging from what nobility in the German Confederation were willing to offer.

What had been lacking up until 1840 was a cohesive national consciousness of what the social problems were and what was needed to resolve them. A national-level consciousness was required to create the national-level cohesion necessary to risk a mass protest against the existing political institution, and a coherent mass protest required leadership. In 1840, the Prussian King Frederick William III died, and his son Frederick William IV, who assumed the throne, would unintentionally provide that cohesion and national leadership. It was also in 1840 that the Rhine Crisis came about when France, having ultimately been humiliated in July 1840 by Russia, Prussia, Austria, and England in its bid to expand influence into North Africa, reasserted its territorial claim to the Rhine River as an internal politically face-saving measure.

Past historical evaluations of the Rhine Crisis of 1840 have indicated that this was a significant event which led to increased German nationalism and allusions of unification (Vanchena, 2000). More recent evaluations of this event indicate that there was little, if any, impact on an already preexisting German nationalism, particularly in the Rhineland (Sedivy, 2016; Brophy, 2013). Instead of nationalism, what Karl Marx believed created a unifying force within Germany beginning 1840 was the growing momentum toward industrialization. He argued that industrialization and its resultant trade required a national system to support it, something that was nearly impossible under the existing fragmented structure called the German Confederation (Marx, 1851, October 25, p. 6).

It is likely that the bourgeoisie were growing restive prior to 1840, based on the expanding trade agreements between states, particularly via Prussia under King Frederick William III. With his death, and his son assuming the throne, who was believed to be more progressive than his father, it may be argued that this belief further promoted thoughts of German unification and the long promised representative constitution. An article published in the *Dusseldorfer Zeitung* in September 1843 clearly defined the same problem that Marx identified in his 1851 article, namely, a fractured system that was unable to support a growing commercial sector (Dusseldorfer Zeitung, 1843).

Social pressure for unification and a representative constitution were building, but instead of fully supporting both, King Frederick William IV attempted to discretely retain a feudal structure while trying to maintain the appearance of being progressive and reducing the social pressure (Hahn, 2001, loc. 1003). Two important Prussian assemblies created by William IV to foster this appearance were the general Evangelical Union in 1846 and the United Diet in 1847. Though recommendations from both assemblies were provided and ultimately ignored by William IV – as they did not support what he wanted – both allowed provincial religious and political leaders to gather at the state level and define the problems that existed within Prussia. As importantly, these deliberations made their way to the national public through the pulpit and the press (Schurz, 1907, p. 72; Howard, 1846, Berlin, October 17), even if what the press wrote had to be read out loud to those who could not afford a newspaper or were unable to read (Sperber, 1994, p. 175).

As background, after Prussia's defeat by Napoleon in 1808, in an effort to promote unity within Prussia, the Prussian Ministry of the Interior assumed what was then to be temporary control of religion within Prussia. This placed schools and churches under the provincial government. In 1817, along with promising an eventual national or general synod, King Frederick William III decreed the Evangelical Union, officially combining Lutheran and Calvinist churches into a state church. In an attempt to further this forced religious unity, King Frederick William III wrote a new confessional policy in 1828 that was to address differences in existing confessional disagreements within the state church. What this lacked was a compromise addressing in any lasting manner the differences that existed between Lutheranism and Calvinism, differences which only the promised national or general synod could have successfully resolved (Jensen, 1974, pp. 142–144; Design Principle 3). As the desire for a representative constitution grew, so did the desire for a national synod to resolve differences within the State Church. With the death of King Frederick William III, there was renewed hope that King Frederick William IV would fulfill the promise of his father. This hope seemed to be validated in 1844 when William IV called up synods at the provincial level within Prussia (Jensen, 1974, pp. 142–144). The same year, Karl Marx, not sharing that hope, wrote of William IV's effort as nothing more than pretending to be progressive while still promoting the old feudal system with himself as its leader (Marx, 1844, p. 4546). In 1846, King Frederick William IV called up a general Evangelical Synod of Prussia, expecting the synod to support his views and his desire to establish an episcopal reorganization on the basis of the Anglican model (Jensen, 1974, p. 149). The British envoy Henry Howard, concerned that discontent from political and religious issues were building on each other, observed this interleaving with regard to the national synod (Howard, 1846, Berlin, June 17). In the end, the King did not obtain the support he desired – the synod was disbanded and their recommendations to the King ignored. Although any disagreements that may have existed between various leaders, clergy, and laity in the Church with the King might not have reached the press, those disagreements certainly carried over into the pulpits of Prussia and beyond (Howard, 1846, Berlin, October 17). Wilhelm Weitling, formally a tailor journeyman-turned activist author, was well received by the working class from which he had his roots (Hahn, 2001, loc. 1102). Shortly after parting ways with Karl Marx, he summed up the religious and political connection in 1847 by stating,

> Church and state have agreed rather to instruct the people in belief, than by belief raise them to knowledge. They have progressed so far as to protect religious freedom or free belief, but the freedom on knowledge is led by mammon with a golden line, and every day shortened by it.
>
> *(Weitling, 1847, pp. 13, 66, 76, 97)*

In their own way, both Weitling and Marx had agreed that the state-controlled church was not legitimate, but Weitling rebelled through interpretation of the Bible, showing the state as it existed was corrupt along with the church that supported it. Marx in typical manner just took an editorial bulldozer to both. In their own way though, they both agreed that the common man was just as close to God as nobility, if not closer, given the wealth and egocentric lives of much of the feudal nobility who went so far as to claim their mandate was from God, likely in an effort to justify their position and community status (Tajfel and Turner, 2001, p. 98). These views, and those of a similar nature expressed by so many others up to this point, show the danger of a combined church and state, all leading back to the observation made by Henry Howard. After the failure of the Evangelical Synod to support William IV's wishes, the United Diet was called up a year later by King Frederick William IV in 1847 to again make it appear he was meeting the other promise made by his father William III while hoping it would support his needs, which, as Karl Marx points out, was a loan to the government for supporting the construction of a Prussian railway. The United Diet ultimately refused to grant the loan and, instead, asked King Frederick William IV to fulfill his father's promise (Marx, 1851, October 28, p. 6).

The United Diet had been King Frederick William IV's answer to keeping the promise his father had made in 1813, which was to establish a representative govern-ment. The British envoy to Berlin described this as the intent of the new King in his 18 June 1842 letter (Hamilton, 1842, Berlin, June 18). The United Diet, first held in April 1847, was the Committees of the states (Prussian provinces) by any other another name. In his opening speech, the King explained why he did not grant a con-stitution, indicating it would be unacceptable to Prussia and something he would not condone (Earl of Westmoreland, 1847, Berlin, April 11). He ended his speech noting that if the United Diet proved useful, he would call them together frequently. The United Diet was not just composed of conservative absolutists though, it contained a spectrum of political viewpoints, many of which were dissatisfied by the limited scope of the United Diet. Further dissatisfaction by a majority of the United Diet came about through their having noted that the Prussian ordinance in 1820, signed by the late Prussian king, established a periodical assembly of the General States, hence it was felt it should not be up to King Frederick William IV, now that an assembly was established, to determine when the assembly was to meet and for what purpose (Earl of Westmoreland, 1847, Berlin, April 15). The United Diet finished July 1847. A let-ter from the British envoy in Berlin painted the results of the first sitting as positive, a viewpoint that differs with that of what Karl Marx offered (Marx, 1851, October 28, p. 6), though both agree that the United Diet had not met the King's expectations (Howard, 1847, Berlin, July 14). What is significant is what the Earl of Westmoreland communicated to Viscount Palmerston on 25 June 1847, toward the end of the United Diet's deliberations. In his letter, he noted how the country had followed the delibera-tions, and, as importantly, how the United Diet through its deliberation process created national-level figures for leadership supporting a representative government (Earl of Westmoreland, 1847, Berlin, June 25).

Carl Schurz, then an activist and student at the university in Bonn, looking back in retirement as one of many who emigrated to America after the revolution of 1848, and who subsequently rose to become a United States Senator, noted basically the same thing when he wrote in 1907,

The United Diet could indeed not resolve, but only debate and petition. But that it could debate, and that its debates passed through faithful newspaper reports into the intelligence of the country – that was an innovation of incalculable consequence. The bearing of the United Diet, on the benches of which sat many men of uncommon capacity and liberal principles, was throughout dignified, discreet, and moderate. But the struggle against absolutism began instantly, and the people followed it with constantly increasing interest.

(Schurz, 1907, p. 73)

Carl Schurz further captures the mood in Bonn during the sitting of the United Diet by noting that everyone listened to the words from Diet members such as Camphausen, Vincke, Beckerath, and Hansemann. Many of these United Diet members would reappear a year later during the revolution as members of the National Assembly. To create the mood, Karl Marx noted in 1851 that the bourgeoisie via the United Diet had shown a rupture with the Government and that a revolution by the bourgeoisie was impending (Marx, 1851, October 28, p. 6). Much of the bourgeoisie and middle-class now had a common understanding through dialogue of what was needed and through the Evangelical Synod and United Diet, how a representative constitutional government could function to meet those needs. With increasing social pressure for unification, freedom of the press, trial by jury, and a liberal constitution, the bourgeoisie and middle-class only needed a final coalescing event, one that would provide the social cohesion to support a reaction. This can also be said for the lower-class workers and peasants, but the social pressure from their particular social needs and wants began to increase starting 1845 with the first bad crop year.

There had been food riots and disturbances in the past, but it was the 1840s which constituted the predominance of food riots within the German Confederation, having reached a climax in 1847 – the same year the United Diet was first held – with a total of about 200 disturbances involving a total of 100,000 participants (Gailus, 1994, p. 172). The increasing number of food-related riots is attributed to the accelerating cost of wheat beginning in mid-1845, reaching a peak mid-1847, after which a good harvest that year began to reduce the cost of wheat, household expenditures (Berger and Spoerer, 2001, p. 303), and hence social pressure in the lower-classes. The largest number of food riots that year occurred in the Confederation States of Prussia (fewer in the western provinces of Rhineland and Westphalia), Bavaria, and Wurttemberg. In Berlin for example, 11 days after the opening of the United Diet, a food riot occurred. To quell the unrest, the military was brought out armed to patrol the streets the remainder of the night. Out of this incident, about 300 people were arrested with 107 of those taken to court (Gailus, 1994, p. 172). The British envoy in Vienna stated in a June 1847 letter that the population in general, to include many of high nobility, men of letters, the learned professions, and the Burghers, appeared discontented with the government as it was (Lord Ponsonby, 1847, Vienna, June 15). From this, it is apparent there are many citizens of various classes within the German Confederation whose social wants and needs are not being met. Those social wants and needs differed by class, as did the pressure to resolve them, but at the present social state, it would only take a common catalyst which all subgroups (i.e., classes other than nobility) with unmet social wants and needs could identify with, to create the proper conditions for not only a riot or revolt, but also a revolution, given the current repressive regime, poor economic conditions, and past promises for a representative constitution.

7.1.3 Amplification of Social Dissonance

From an economic point of view, numerous manufacturing firms failed during the years 1846 to 1848, with lack of credit being cited as the main cause. For example, investment in railways was reduced, seriously impacting the metals and mining industry. Based on their economic analysis, Berger and Spoerer conclude that the agrarian crisis led to an emerging industrial crisis leading into 1848 (Berger and Spoerer, 2001, p. 306). Though the economic analysis is excellent and the data persuasive, it is argued here that the industrial crisis, or manufacturing shock as they label it, was just one more variable adding to increasing pressure for reaction and change within the German Confederation. Ultimately, it is the observations made by German citizens and British envoys at the time and immediately after the February 1848 French revolution that will be used in determining what they believe to have initiated the German Confederation's revolution of 1848. Before addressing that aspect though, one final variable is introduced for consideration – that of emigration and the amplification of discontent it may have had on the general population of the German Confederation during the bad crop years of 1845 to 1847.

Related to revolution is emigration, where emigration is defined here as an individual's reaction based on the magnitude of unmet social wants and needs, cost and risk to emigrate versus the cost, risk, and likelihood of national pressure that leads to successful political institutional improvements (once again Axiom 2 appears in this). As alluded to in previous chapters, this is equivalent to changing queues if the original queue is moving so slowly or is so dysfunctional that a member feels it is unlikely that his or her needs will not be met within the necessary time available. Beyond just the individual, increasing emigration (changing queues) signals to those who remain in country (in the original queue) that there is increasing general discontentment with the political system (resource system). In effect, this creates awareness of shared grievances motivating reaction by those who remain – leading potentially to insurrection or revolution assuming a sufficient social pressure and density of discontented citizens who remain in country exist (Pfaff and Kim, 2003, pp. 406–409). Emigration data, calculated for the decade leading up to the German Confederation's revolution in 1848, is shown in Figure 7.1. It is estimated that about 90 percent of emigrants from the German Confederation emigrated to the United States during this period (Burgdorfer, 1931, p. 332). Viewing Figure 7.1, the emigration rate from 1825 to 1831 indicates either high emigration risk or low unmet social wants and needs, possibly due to good crop years and low food prices most of these years (Burgdorfer, 1931, p. 342), when compared to data from 1832 to 1844. Because of the commercial crisis in the United States starting in 1837, the risk of emigrating to America increases significantly, implying the emigration rate would likely have been higher otherwise. Once the European crop failure began in 1845 and the commercial crisis had ended in

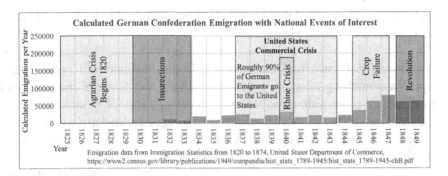

FIGURE 7.1 German Emigration by Year, Beginning 1825 through 1849, with Major Historical Events for Possible Correlation with Emigration Rate.

the United States, the yearly emigration rate increased until by 1847 it had quadrupled as compared to the average rate from 1832 through 1844. In just 3 years, 1845 through 1847, about one-half percent of the entire population of the German Confederation had emigrated. The reduced emigration rate beginning in 1848 with the revolution might be interpreted as a result of now decreasing food prices, and/or possibly resulting from national anticipation of political improvement, maybe even the long promised representative constitution and free press. What is important is that the recent and significantly increased emigration rate that peaked in 1847, when combined with the nationally published discussions out of the United Diet, also in 1847, created a common awareness of discontent. A shared social tension was now in place, a feeling that something was needed to address the social needs and wants identified by emigrating neighbors and representative citizen members of the United Diet. All of this added to the uncertainty information on which the community would validate its belief of what should be, and then base its decision regarding whether to react or not react based on what currently is.

Beyond emigration as a means to amplify social dissonance related to the German Confederation's failure of governance are the underground clubs and various social groups at the time. As Festinger (1957, p. 177) has noted, interaction within a member's appropriate social group (since there are usually many) can be an important means for increasing or reducing social dissonance. In doing so, the group is confirming the beliefs of the particular member. This and supporting public debate all contribute to reducing the social noise (N_0) whereas the risk involved adds to that noise. With risk and public opinion now well established within the German Confederation, what is necessary for a reaction to occur at this point is just one final social push, or amplification, from a neighboring country and its citizens that many within the German Confederation identified with as a group member.

7.1.4 The Amplifying Event and Group Member Reaction

Word of France's revolution reached Bonn university student Carl Schurz near the end of February 1848 as he was studying in his upstairs room. He recalls that upon hearing the news, he ran down the stairs to the market-square in Bonn where the student societies usually gathered, and that day they quickly gathered to understand what the revolution meant for them. He states,

> But since the French had driven away Louis Philippe and proclaimed the republic, something of course must happen here, too. Some of the students had brought their rapiers along, as if it were necessary at once to make an attack or to defend ourselves. We were dominated by a vague feeling as if a great outbreak of elemental forces had begun, as if an earthquake was impending of which we had felt the first shock, and we instinctively crowded together.
>
> *(Schurz, 1907, p. 75)*

The next day, as they further spoke among themselves and with strangers they met on the street, he recalls that they concluded the day had arrived for, "establishment of 'German Unity' and the founding of a great, powerful national German Empire" (Schurz, 1907, p. 76). In a similar initially defensive manner, after word reached leadership within the German Confederation that a revolution in France had taken place on 24 February 1848, the Federal Diet convened in Frankfurt to consider what defensive actions the Confederation should take in case France were to attack its borders along the Rhine.

This reaction by the Federal Diet was a precautionary measure, but given the relatively recent experience with Napoleon and then the Rhine Crisis, it was a response that can easily be appreciated. Beyond just the fear of attack, Frederick Orme, the British envoy in Frankfurt, noted in his 2 March 1848 letter the impact the French Revolution was having on Germany and in particular the southwestern states that more readily identify with France (Orme, 1848, Frankfurt, March 2). This observation may have been influenced by at least one petition that had already been submitted, as Friedrich Hecker recalled, with demands from the citizens and residents of Mannheim to the Second Chamber (Constitutional Charter, 22 August 1818) of the Baden Assembly on 26 February 1848. The petition noted that France had just been transformed by its revolution and that the German people had the right to demand prosperity to include education and freedom for all classes of society, regardless of birth or class. It went on to state that the time for long deliberations was past, what the people wanted had already gone through the press and proper legal representatives (i.e., implying it was fit to send as a petition to the Second Chamber), and that measures for adoption needing emphasized were

1) Creating a citizen's militia and with free elections for its officers.
2) Absolute freedom of the press.
3) Jury courts modeled after those in England.
4) Immediate creation of a German parliament.

(Hecker, 1848, p. 18)

After realizing by 10 March 1848 that it was unlikely France would attack the Confederation, and that the social reaction instigating change had become irresistible, the Federal Diet began deliberating in secret on the establishment of the long-promised German national parliament (Orme, 1848, Frankfurt, March 10), a concept that would quickly evolve into what would become the National Assembly in Frankfurt.

Finally, as one further attestation as to what triggered the March 1848 revolution in the German Confederation, in his 28 October 1851 New-York Daily Tribune article, Karl Marx states,

> In Prussia, the Bourgeoisie had been already involved in actual struggles with the Government, a rupture had been the result of the "United Diet;" a Bourgeoisie revolution was impending, and that revolution might have been, in its first outbreak, quite as unanimous as that of Vienna, had it not been for the Paris revolution of February. – That event precipitated everything
>
> *(Marx, 1851, October 28)*

Carl Schurz; Dr. Friedrich Hecker and his associates; the Federal Diet; the British envoys in Frankfurt, Berlin (Earl of Westmoreland, 1848, Berlin, February 29), Hamburg (Hodges, 1848, Hamburg, March 6), Dresden (Forbes, 1848, Dresden, March 4), Hanover (Bligh, 1848, Hanover, March 24), Vienna (Viscount Ponsonby, 1848, Vienna, March 1), Stuttgart (Malet, 1848, Stuttgart, March 2), and Munich (Milbanke, 1848, Munich, March 4); and Karl Marx, all attributed the March revolution within the German Confederation to the French revolution. The earlier crop failure, pressures on the guilds created by a growing industry, the industrial crisis, and a dysfunctional form of governance certainly would have increased the social needs and wants of Confederation members – but heightened social needs and wants do not start a revolution, they are merely a requirement. Reaction in the form of revolution arising from increasingly unmet social needs and wants requires a national-level vision of necessary political institutional improvement based on social identity and norms (preferred traits and common beliefs) which must evolve and be accepted over time. Not a class vision, nor just an individual's vision of what could be perceived as a chance for improvement, but a national vision – all at a level of risk the participating citizens were willing to accept.

The basic stage had been set for revolution within the German Confederation by the end of 1847. Nobility had shown and continued to demonstrate a lack of concern for anything but themselves, and their mandate from God was viewed with decreasing authenticity if it ever had any to begin with. The citizens of the German Confederation had been promised a representative constitution at the inception of the Confederation by King Frederick William III, only to be teased with the concept by his son who had his own personal agenda which was demonstrated to not involve the creation of a representative government. The insurrections of 1830–1833 created a collective memory of the Paris Revolution as having led to local insurrections in the German Confederation, so revolution after the February 1848 French revolution could be considered and justified based on the memory of those who had been young at the time but who were older and in positions of influence by 1848 (Pennebaker and Banasik, 1997). The general Evangelical Synod of 1846 and the United Diet of 1847, both held within Prussia, provided a Confederation-level vision of what could be achieved while also showing that dissention was allowed (i.e., acceptable risk). Finally, the food riots of 1847 demonstrated to the lower-class how their voices could be heard and that breaking of the law offered risk, but less risk than posed by possible starvation at the hands of an incompetent government. In the end, France revolted against its King on 24

February 1848, finally pushing citizens within the German Confederation to choose between the national identity they desired to be part of or continue to serve under a government that did not represent who they were or address what they wanted. The revolution eventually failed, but the community-wide reaction is what was of interest. In effect, the experimenter of Stanley Milgram's obedience experiments was disobeyed (Milgram, 1965), and citizens of the German Confederation revolted based on their French European group members they could better identify with politically.

7.2 Response to Intrusion Into Waiting Lines and the German Confederation's Nobility

As Friedrich von Coelln had recognized on or before 1807, nobility had become a free-riding social subgroup within what was basically a semi-Feudal German Confederation. Their subgroup identity was distinct from the community from which their wealth and power was derived, and a hierarchical structure existed within this subgroup. The community's view of nobility as free-riders persisted from at least 1807 and likely grew leading up to the 1848 revolution as evidenced by Johann August Wirth's commentary in 1832, an article in the Dusseldorfer Zeitung (1843), along with the multiple articles by Karl Marx and others. Viewed within the structure of a queue, free-riding nobility are those that intrude into the queue at the front, offer relatively little or no benefit to the community commensurate with resources which they consume, and thus reduce opportunity (resources) for the other queue members already in queue who are not of noble blood. As the various offspring of nobility and their relations proliferate, the problem is further exacerbated due to established social norms of the nobility which lead to placing their relations in positions of influence, which in turn increases the priority intrusion rate for rival resources. If the intrusion is infrequent, then tempers of those impacted within the lower-class community flare, then pass away over time. If such events become so frequent so as to impact the ability of those who have gained status in the lower class and have been waiting in queue to gain access to the desired resource (food, housing, education, wealth, power, etc.), then the intrusions and resultant dissonance begin to accumulate, possibly reaching equilibrium at a still acceptable level of discomfort, or at the extreme, ending in leading or supporting revolution as the final solution if the emotional discomfort and associated pressure to react reach sufficient magnitude.

As a generic example, consider a hypothetical situation where, every day, a citizen reads in the newspaper or discusses in his or her club how nobility is committing some variety of yet another social deviation. To make this simple, assume that various deviations identified each day have the same stimulus intensity, say one unit, and associated unit sensation magnitude of 1. If the meantime to decay for the various social deviations is also assumed the same for each stimulus type, say $\lambda = 3$ $days^{-1}$, then the resultant stimulus intensity approaches equilibrium which, using Equation 6.1, is,

$$e^{-\frac{1}{3\,days}\cdot\left(t_{final}-t_{final}\right)} + e^{-\frac{1}{3\,days}\cdot\left(t_{final}-\left[t_{final}-1\right]\right)} + \ldots + e^{-\frac{1}{3\,days}\cdot\left(t_{final}-1\right)} \cong 3.528$$

$$\rightarrow I_n(t)\,over\,time\,as\,both\,n,t \rightarrow \infty.$$

Under the given simplifying assumptions, and assuming the group members to which this information is directed listen to nothing else, this results in a unit sensation magnitude of,

$$\overline{es}_{3.528} = ln\left(\frac{3.528 + \widehat{N}_0}{\widehat{N}_0}\right).$$

Social cohesion also plays a role, as do the number of group members in a hierarchical system who amplify the sensation magnitude of their leaders. But the example is clear in that all that is needed for leadership to increase the probability of reaction by community members is to accurately monitor on a regular and frequent basis the social deviations as perceived within the community member social subgroup they represent. This is an argument for the importance of Design Principle 4 and factual unbiased monitoring of events within the community. When monitoring panders to the belief of a subgroup in the community or the viewpoint desired at the time by those of higher status whose concern is to maintain or increase their status, then community resources are at risk. It is this reason that dictators, autocrats, and the like resent a free press and do what they can to nullify it, beginning with calling it fake and making use of the uncertainty information effect, particularly through non-reaction $UI = \{NR_1\}$, which, as Chapter 3 has shown, is more effective at reducing probability of reaction than $UI = \{R_1\}$.

The German Confederation's nobility isolated themselves from the people who financially supported them. Nobility legally avoided taxes. Those of nobility often had mistresses and associated scandals in an otherwise religious community, and they were often on vacations at expensive resorts that the general populace could not imagine and could only hear rumors of, and yet nobility did nothing to benefit the community. These and other transgressions were counter to the beliefs of their citizens and, when continually brought to light by a free or at least semi-free press allowed for an equilibrium of stimulus intensity to develop for that stimulus (e.g., free-riding nobility and dysfunctional governance). From this, it is possible that cognitive dissonance from various associated stimuli, supporting in some way the need for unification and general public representation, was additive. This is where social space becomes complicated, possibly leading into a multidimensional vector field, and where much work remains in developing the theoretical underpinnings for future and more comprehensive analyses.

7.3 The Obedience Experiment and the German Confederation's Revolution of 1848

From Chapter 4, recall the Condition 4 obedience experiment where two confederates sequentially disobey the experimenter (Milgram, 1965). In this particular experiment, the first confederate disobeys the experimenter when the naive subject (teacher) reaches 150 volts, thus prompting over 25 percent of the naive subjects out of the 40 trials conducted to disobey at that voltage or shortly after. In its most simplistic form, the German Confederation beginning March 1848 may be equated with the naive subject at 150 volts, and the revolution in France beginning February 1848 as the first confederate defying the experimenter. It is argued that the political festivals, newspaper editorials, expanding industrialization, the General Synod, the United Diet, and emigration all helped enforce beliefs and reduce social noise so that the populace of the German Confederation better understood why the previously promised representative government was needed. Thus, a concept introduced in 1813 under King William III solidified into a belief by 1847 of what should be which led to the associated dissonance caused by a divergence between belief of what should be and what was. The means for establishing a representative government had been demonstrated through the United Diet; now all that was needed was an added push to react, or, in this case, a confederate group member supporting a reaction based on the belief of what should be done, given the social norms (belief) of the German community.

As indicated, the stimulus intensity for unification developed over time, but amplification of the associated sensation magnitude within the German Confederation is argued to have occurred immediately before the Confederation's revolution. Switzerland had just had a minor civil war leading to improved governance in late 1847, and Italy was in the process of revolution when France began its 1848 revolution. This implies the possibility of three examples of countries in revolt, with France having the greatest commonality of factors and identity with members of the German Confederation, particularly its southwestern states. Assume that the concept of cross-modality matching (Stevens, 1975, pp. 33–34) may be applied in this case between a queue intrusion stimulus and the stimulus caused by free-riding of the nobility, then if unit sensation magnitude comparable to an illegitimate intrusion may be argued to exist, such that $\overline{es} \cong 1.148$, then to achieve a greater than 50 percent probability of reaction when just considering France as a confederate model, then noise $\widehat{N}_0$ would have to be less than or equal to 2.22. From this, if $\widehat{N}_0 \leq 2.22$ and the potential cumulative effect on stimulus intensity of social deviations by the nobility over time is not yet accounted for, the general model proposed for probability of revolution could be bounded by,

$$\left[1 - e^{-1.148 \cdot ln\left(\frac{1+2.22}{2.22}\right) \cdot ln\left(e \cdot \frac{1+1.147}{1.147}\right)} \right] =$$

$$0.5 \leq P\left(Revolution \mid French\ Revolution\right) \leq 0.846$$

$$= \left[1 - e^{-1.148 \cdot ln\left(\frac{1+0.582}{0.582}\right) \cdot ln\left(e \cdot \frac{1+1.147}{1.147}\right)} \right] for\ \widehat{N}_0 = \frac{1}{e-1} \cong 0.582.$$

Where $\widehat{N}_0 \cong \frac{1}{e-1} = 0.582$ might be approached if social noise had been reduced to the minimum due to festivals, via reporting by the semi-free press, the General Synod, the United Diet, emigration, and lack of strong disincentives (risk) from the Monarchy. For comparison purposes, the noise level for a Grand Central station queue was found to be about $\widehat{N}_0 = 2.056$. Given the uncertainty involved and the limited risk of confrontation, there would seem to be comparisons, with the possibility that social noise was lower for most of the German citizens in 1848 than for those New Yorkers in 1978 waiting in queue. A 50/50 chance for revolution is not good odds, and anything above that is just a reaction waiting to happen. Sans revolution though, it would also be interesting to examine the relationship between lower sensation magnitude levels and emigration.

7.4 Final Remarks

Design principles 2 (earned status) and 4 (unbiased monitoring by the community and free-riding) were clearly violated from at least 1807 to 1848 in what was to become, and what became the German Confederation beginning in 1815. Given that the monitoring of nobility continued to occur even when the free-press was repressed, that nobility had increasingly become free-riders, and that the system of governance did not address the wants and needs of the general populace, the German Confederation, being a community with clearly defined boundaries, was a commons doomed to eventual failure. It was an outdated, outworn institution based on the feudal structure, which in the end was unable to satisfy the needs of its citizens as they in turn observed and tried to compete with the emerging industrial nations around them. Yet, the nobility resisted necessary changes to the political and social structure. They were wealthy, they had power, the resource system was working well for them, and as a means to reduce dissonance, as a group with its own identity, they likely convinced themselves that this is the way it should be. A decrease in status for them would appear to cause greater emotional discomfort than dealing with occasional insurrections, and as per Axiom 2, nobility would thus be more than willing to accept the occasional insurrection.

Throughout history, the question arises as to how rich is rich enough, and if a person is working, why should they be poor and barely able to survive? German workers in the southwestern region of the Confederation were asking and debating this question in 1848 (Esselen, 1848, pp. 37–40), and the question is being asked in many countries around the world now. Typically, wealth brings along power and status from which the wealthy may then influence legislation and social perception to maintain or increase their wealth via the media and political representatives they control. But to what end? Is it a competition to be the wealthiest person alive at the expense of those who are getting the work done to create that wealth? It is fair that the wealthy gain their wealth through the use of community-purchased infrastructure without paying sufficient contributions supporting the very same infrastructure – the educational infrastructure being a prime example. It may also be asked if the governance structure is distributing tax revenue in a way that best benefits the long-term welfare of the community as a whole. A systems level approach needs to be taken when addressing social needs and wants of the community, else resources are not optimally used. To enable this approach, the community must participate, it must be educated, and it must be interested enough to pursue these questions and force necessary change when required, since those relying on the status quo to remain in power do not want to lose their status (position in queue) as that creates dissonance for them.

From a more fundamental view, we live in a closed system, and there are only so many resources on which to base our survival. As with cattle, if too many exist in a closed pasture relying on a rival resource, they will all starve. Going to another planet (pasture), when eventually possible, might work for a few, but not for the masses left behind, and in the end, if we do not learn to live properly on this planet, then we will just go to another and start the cycle over again. A solution needs to be found by all communities on this planet, and that is where we all must find common beliefs leading to a common goal. As King William III discovered, dictating a belief to the citizens is not effective. Identifying common beliefs though relies on common and accurate information, and that is the difficulty we face in an age of instant communication with varying levels of reliability and access by community members. Community members may currently select the information they want to hear in order to reduce their own dissonance, or the information they hear is limited or slanted based on what those in power want them to hear.

Accurate information heard by all enables citizens of a commons, with adequate education, to understand the problem and reach a consensus on the rules they want to follow for maintaining stability of their resource system. If they are not able to reach consensus, then stability will not be achieved. In all cases, the basic problem is that all citizens must see an issue of interest as a social deviation, and they must understand and appreciate its relevance. Otherwise, we as humans tend to react only when the problem is hitting us directly across the face, and that leads us back to the need for improved education which in turn allows us to perceive past patterns as they relate to current events, appreciate science as it should be practiced, read about and debate our differences, and, as a result, better focus on questions that reduce social noise and pursue reactions as a community that more effectively addresses demanding issues. With escalating social alterations due to climate change, more frequent pandemics due to increasing population density, the information explosion resulting in a contest where we either make use of it or are used by it, population shifts from wars and famines, and as their impact becomes more relevant to all in our shared social commons, it seems an appropriate time to consider a more participatory and equitable form of governance, a stable form that mitigates disruptive and often horrific oscillations between the various forms of governance which have been experienced to date. To do this, though, we need to understand and address what seems to be a basic human need for status and how best to identify and reward status to those deserving of it, based on how they benefit the community and how to monitor and swiftly remove that status when benefit to the community is no longer provided.

We have history as data. Lessons learned from the past do not have to be repeated if they are actually learned from and passed on in a meaningful manner. Lessons learned can also involve approaches to social governance that worked well for a time, with the need then to understand what eventually led to their failure. Three examples for consideration to start the process are from the Old Testament (NIV Bible, 2011, 1 Samuel 8), Imperial China's civil service examination system which promoted some amount of equity, and the ancient democracy of Greece (Ober, 2015). It is not enough to learn of and debate historical events, which is the process for data collection and validation, but the data must be placed within a framework supporting systematic analysis of complex social structures to clarify why events unfolded as they did. With cause and effect identified, patterns emerge, and patterns lead to further quantitative modeling. Mistakes will be made in developing such a process, mistakes have probably been made in this book, but that is how we learn and grow. Up to this point, and one of the reasons for expending effort to write this is that we seem to be stagnating from a social development standpoint. Maybe Sorokin (1985) was correct in his cyclic view of societies, but we need to eventually break the cycle if it does exist in order to grow as a world community.

Basic mathematical tools and social concepts have been proposed here that may potentially allow expansion into more socially complex areas of social space such as those described before. By expanding the boundaries of social space, further mathematical development will ensue. Use history as social data. Use it to further develop and validate tools and theory presented to this point in an effort to systematically expand on their application. This is where collaboration across academic disciplines is critical since the limited view and capabilities of a few clearly limit the speed of progress as compared to what is deemed necessary, given our current global social situation. The end goal of this effort should be to determine if there is a way to achieve and sustain a stable social commons (i.e., stable node) on this planet other than relying on another 100,000 years of human evolution. With that, let us all make a sustainable and peaceful social commons our goal, as the alternative, based on history, is never pleasant.

References

Bavarian thaler. (1829). *Commercial Treaty*. Retrieved from www.pcgs.com/cert/37715472

Berger, H., & Spoerer, M. (2001). Economic Crises and the European Revolutions of 1848. *The Journal of Economic History*, 61(2), 293–326. https://doi.org/10.1017/S0022050701028029

Bligh, J.D. (1848, Hanover, March 24). Letter to Viscount Palmerston. In *British Envoys to Germany, 1816–1866: Volume III 1848–1850*. Cambridge University Press, 2006.

Brophy, J. (2013). The Rhine Crisis of 1840 and German Nationalism: Chauvinism, Skepticism, and Regional Reception. *The Journal of Modern History*, 85(1), 1–35. https://doi.org/10.1086/668802

Burgdorfer, F. (1931). Migration Across the Frontiers of Germany. In W. Wilcox (Ed.), *International Migrations: Vol II* (p. 342). National Bureau of Economic Research; Immigration Statistics from 1820 to 1874, United States Department of State. Retrieved from www2.census.gov/library/publications/1949/compendia/hist_stats_1789-1945/hist_stats_1789-1945-chB.pdf

Cartwright, T. (1831, Frankfurt, November 23). Letter to Viscount Palmerston, In *British Envoys to Germany 1816–1866: Volume II 1830–1847*. Cambridge University Press, 2002.

Cartwright, T. (1832, Frankfurt, June 25). Letter to Viscount Palmerston. In *British Envoys to Germany 1816–1866: Volume II 1830–1847*. Cambridge University Press, 2002.

Cartwright, T. (1832, Frankfurt, July 16). Letter to Viscount Palmerston. In *British Envoys to Germany 1816–1866: Volume II 1830–1847*. Cambridge University Press, 2002.

Cartwright, T. (1833, Frankfurt, April 5). Letter to Viscount Palmerston. In *British Envoys to Germany 1816–1866: Volume II 1830–1847*. Cambridge University Press, 2002.

Chad, G. (1830, Frankfurt, September 9). Letter to George Hamilton-Gordon. In *British Envoys to Germany 1816–1866: Volume II 1830–1847*. Cambridge University Press, 2002.

Chad, G. (1830, Frankfurt, September 11 and 22). Letter to George Hamilton-Gordon. In *British Envoys to Germany 1816–1866: Volume II 1830–1847*. Cambridge University Press, 2002.

Coelln, F. (1807). *Neue Feuerbrande*. Amsterdam and Colln.

Constitutional Charter. (1818, August 22). *Constitutional Charter for the Grand Duchy of Baden*. Retrieved from https://germanhistorydocs.ghi-dc.org/sub_doclist.cfm?sub_id=14&section_id=9

Deutsche Bundesakte. (1815, Juni 8). (Jeremiah Riemer, Trans.). Retrieved from https://germanhistorydocs.ghi-dc.org/sub_doclist.cfm?sub_id=14&section_id=9

Dusseldorfer Zeitung. (1843). *Germany's Unification* (Jeremiah Riemer, Trans.). Dusseldorfer Zeitung, No. 244 and 246, 3 and 5 September 1843. Retrieved from https://germanhistorydocs.ghi-dc.org/sub_doclist.cfm?sub_id=14&section_id=9

Earl of Westmoreland. (1847, Berlin, April 11). Letter to Viscount Palmerston. In *British Envoys to Germany 1816–1866: Volume II 1830–1847*. Cambridge University Press, 2002.

Earl of Westmoreland. (1847, Berlin, April 15). Letter to Viscount Palmerston. In *British Envoys to Germany 1816–1866: Volume II 1830–1847*. Cambridge University Press, 2002.

Earl of Westmoreland. (1847, Berlin, June 25). *British Envoys to Germany 1816–1866: Volume II 1830–1847*. Cambridge University Press, 2002.

Earl of Westmoreland. (1848, Berlin, February 29). Letter to Viscount Palmerston. In *British Envoys to Germany, 1816–1866: Volume III 1848–1850*. Cambridge University Press, 2006.

Esselen, C. (1848, Juni 10). Ein politisches Glaubensbekenntnisz. *Allgemeine Arbeiter-Zeitung*, 33–40. Retrieved from https://sammlungen.ub.uni-frankfurt.de/periodika/periodical/titleinfo/9183299

Festinger, L. (1957). *A Theory of Cognitive Dissonance*. Stanford, CA: Stanford University Press. Retrieved from http://www.sup.org/books/title/?id=3850

Forbes, F.R. (1848, Dresden, March 4). Letter to Viscount Palmerston. In *British Envoys to Germany, 1816–1866: Volume III 1848–1850*. Cambridge University Press, 2006.

Frederiksen, D. (1894). Mortgage Banking in Germany. *The Quarterly Journal of Economics*, *9*(1), 47–76. https://doi.org/10.2307/1883634

Gagliardo, J. (1969). *From Pariah to Patriot*. The University Press of Kentucky (Kindle Edition).

Gailus, M. (1994). Food Riots in Germany in the Late 1840s. *Oxford University Press on Behalf of the Past and Present Society*, *145*(1), 157–193. https://doi.org/10.1093/past/145.1.157

Hahn, H. (2001). *The 1848 Revolutions in German-Speaking Europe*. Taylor & Francis (Kindle Edition). https://doi.org/10.4324/9781315839295

Hamilton, G. (1842, Berlin, June 18). Letter to Earl of Aberdeen. In *British Envoys to Germany 1816–1866: Volume II 1830–1847*. Cambridge University Press, 2002.

Hecker, F. (1848). *Die Erhebung des Volkes in Baden für die deutsche Republik im Frühjahr 1848*. Basel. Retrieved from www.leo-bw.de/web/guest/detail/-/Detail/details/DOKUMENT/blb_digitalisate/169367/Erhebung+des+Volkes+in+Baden+f%C3%BCr+die+deutsche+Republik+im+Fr%C3%BChjahr+1848

Hodges, G.F. (1848, Hamburg, March 6). Letter to Viscount Palmerston. In *British Envoys to Germany, 1816–1866: Volume III 1848–1850*. Cambridge University Press, 2006.

Howard, H. (1846, Berlin, June 17). Letter to Earl of Aberdeen. In *British Envoys to Germany 1816–1866: Volume II 1830–1847*. Cambridge University Press.

Howard, H. (1846, Berlin, October 17). Letter to Viscount Palmerston. In *British Envoys to Germany 1816–1866: Volume II 1830–1847*. Cambridge University Press.

Howard, H. (1847, Berlin, July 14). Letter to Viscount Palmerston. In *British Envoys to Germany 1816–1866: Volume II 1830–1847*. Cambridge University Press, 2002.

Jensen, G. (1974). Official Reform in Vormärz Prussia: The Ecclesiastical Dimension. *Central European History*, 7(2), 137–158. Retrieved from www.jstor.org/stable/4545701

Keynes, J.M. (1919). *The Economic Consequences of Peace*. Liberty Fund, Inc. (Kindle Edition). Retrieved from https://oll.libertyfund.org/title/keynes-the-economic-consequences-of-the-peace

Lord Ponsonby (1847, Vienna, June 15). Letter to Lord Palmerston. In *British Envoys to Germany 1816–1866: Volume II 1830–1847*. Cambridge University Press, 2002.

Malet, A. (1848, Stuttgart, March 2). Letter to Viscount Palmerston. In *British Envoys to Germany, 1816–1866: Volume III 1848–1850*. Cambridge University Press, 2006.

Marx, K. (1844). Critique of Hegel's Philosophy of Right. In *Collected Works of Georg Wilhelm Friedrich Hegel* (Henry Stenning, Trans.). Delphi Classics, 2019 (Kindle Edition), (Henry Stenning, Trans.), first printed in the Deutsch-Französische Jahrbücher, 7 & 10 February 1844 in Paris.

Marx, K. (1851, October 25). Germany: Revolution and Counter-Revolution. *New-York Daily Tribune, 11*(3283), 6. Retrieved from https://chroniclingamerica.loc.gov/lccn/sn83030213/

Marx, K. (1851, October 28). Germany: Revolution and Counter-Revolution. *New-York Daily Tribune*, 11(3285), 6. Retrieved from https://chroniclingamerica.loc.gov/lccn/sn83030213/

Marx, K. (1852). The Eighteenth Brumaire of Louis Bonaparte. *Die Revolution*. Published in New York 1852. Retrieved from www.marxists.org/archive/marx/works/1852/18th-brumaire/

Marx, K. (1912). *Revolution and Counter-Revolution or Germany in 1848* (E. M. Aveling, Ed.). Charles Kerr & Company.

Milbanke, J.R. (1828, Frankfurt, March 14). Letter to the Earl of Dudley. In *British Envoys to Germany, 1816–1866: Volume I 1816–1829*. Cambridge University Press, 2000.

Milbanke, J.R. (1830, Frankfurt, September 29). Letter to the Earl of Aberdeen. *British Envoys to Germany 1816–1866: Volume II 1830–1847*. Cambridge University Press, 2000.

Milbanke, J.R. (1848, Munich, March 4). Letter to Viscount Palmerston. In *British Envoys to Germany, 1816–1866: Volume III 1848–1850*. Cambridge University Press, 2006.

Milgram, S. (1965). Some Conditions of Obedience and Disobedience to Authority. *Human Relations*, 18(1), 57–76. https://doi.org/10.1177/001872676501800105

NIV Bible. (2011). *New International Version (NIV) Bible*. Online. Retrieved from www.biblegateway.com/passage/?search=1%20Samuel%208&version=NIV

Ober, J. (2015). *The Rise and Fall of Classical Greece*. Princeton University Press. www.jstor.org/stable/j.ctt13x0q7b

Orme, F.D. (1848, Frankfurt, March 2). Letter to Viscount Palmerston. In *British Envoys to Germany, 1816–1866: Volume III 1848–1850*. Cambridge University Press, 2006.

Orme, F.D. (1848, Frankfurt, March 10). Letter to Viscount Palmerston. In *British Envoys to Germany, 1816–1866: Volume III 1848–1850*. Cambridge University Press, 2006.

Ostrom, E. (1990). *Governing the Commons*. Cambridge University Press. https://doi.org/10.1017/CBO9781316423936

Pennebaker, J., & Banasik, B. (1997). On the Creation and Maintenance of Collective Memories: History as Social Psychology. In J. Pennebaker, D. Paez, & B. Rime (Eds.), *Collective Memory of Political Events*. Lawrence Erlbaum Associates, Inc. https://doi.org/10.4324/9780203774427

Pfaff, S., & Kim, H. (2003). Exit-Voice Dynamics in Collective Action: An Analysis of Emigration and Protest in the East German Revolution. *American Journal of Sociology, 109*(2), 401–444. https://doi.org/10.1086/378342

Schurz, C. (1907). *The Reminiscences of Carl Schurz*. S.S. McClure Co. (Kindle Edition).

Sedivy, M. (2016). The Austrian Empire, German Nationalism, and the Rhine Crisis of 1840. *Austrian History Yearbook*, *47*, 15–36. https://doi.org/10.1017/S0067237816000059

Six and Ten Articles. (1832). The Six Articles (June 28, 1832) and the Ten Articles (July 5, 1832). (Jeremiah Riemer, Trans.). Protokolle der Bundesversammlung 1832, 24. Sitzung. Retrieved from https://germanhistorydocs.ghi-dc.org/sub_doclist.cfm?sub_id=14&section_id=9

Sorokin, P. (1937/1985). *Social and Cultural Dynamics*. Taylor & Francis. Retrieved from www.routledge.com/Social-and-Cultural-Dynamics-A-Study-of-Change-in-Major-Systems-of-Art/Sorokin/p/book/9780878557875

Sperber, J. (1989). Echoes of the French Revolution in the Rhineland, 1830–1849. *Central European History*, *22*(2), 200–217. https://doi.org/10.1017/S000893890001150X

Sperber, J. (1994). *The European Revolutions, 1848–1851*. Cambridge University Press. https://doi.org/10.1017/CBO9780511817717

Stevens, S.S. (1975). *Psychophysics*. John Wiley and Sons Inc.

Tajfel, H., & Turner, J.C. (2001). An Integrative Theory of Intergroup Conflict. In M.A. Hogg & D. Abrams (Eds.), *Intergroup Relations: Key Readings* (pp. 94–109). Psychology Press, Taylor & Francis.

Vanchena, L. (2000). The Rhine Crisis of 1840: Rheinlider, German Nationalism, and the Masses. In N. Vazsonyi (Ed.), *Searching for Common Ground: Diskurse zur deutschen Identität 1750–1871* (306 pp.). Böhlau-Verlag.

Viscount Ponsonby. (1848, Vienna, March 1). Letter to Viscount Palmerston. In *British Envoys to Germany, 1816–1866: Volume III 1848–1850*. Cambridge University Press, 2006.

Weitling, W. (1847). *The Gospel of the Poor Sinners*. Published by the Author.

William III. (1807/1902). The Prussian Reform Edict (9 October 1907). In J.H. Robinson (Ed.), *Translations and Reprints from the Original Sources of European History*, vol II, no. 2: The Napoleonic Period. University of Pennsylvania, 1902. Retrieved from https://revolution.chnm.org/d/517

Wirth, J. (1832). *Johann August Wirth at the Hambach Festival (May 1832)* (Jeremiah Riemer, Trans.). Das Nationalfest der Deutschen zu Hambach, Neustadt. Retrieved from https://germanhistorydocs.ghi-dc.org/sub_doclist.cfm?sub_id=14&section_id=9

APPENDIX A

Deriving Minimum Queue Length Based on Milgram et al. (1986) Data

The accuracy of a model is only as good as the model and the data from which it is derived. The data presented in Milgram et al. (1986) is what was necessary for their article, and even that limited amount of data is greatly appreciated. Unfortunately, what is really needed are the original coding sheets used by the graduate students to log raw data for each intrusion event. Figure A.1 provides the only known example of a completed coding sheet from the Milgram et al. (1986) intrusion experiment conducted in 1978 by his graduate students.

Figure A.1 information is as follows:

1. Condition 2 Coding Sheet for Team A:

    a. 1 Buffer – Christina Taylor and 1 Intruder – David Nemiroff
    b. Observer Recording Data – Ronna Kabatznick

2. Location: Ticket Line at Grand Central Station, New York City
3. Date and Time: 5 May 1978 at 4:05 p.m.
4. Intrusion Point: Three members ahead (one male, two females), five members behind (two females, three males) including confederate female buffer. Total queue length of eight including confederate buffer.
5. Reactions:

    a. Line position 5/Position (+2) verbal ejection of intruder
    b. Line position 3/Position (−1) verbal disapproval to intruder
    c. Line position 6/Position (+3) verbal disapproval to others – not counted as reaction since not directed at intruder as per Milgram et al. (1986)

6. Qualitative Notes Recorded by Observer:

    a. Line position 5/Position (+2) "Hey buddy, the line forms in the rear."
    b. Line Position 3/Position (−1) "What's this? I'd like to get in here?"

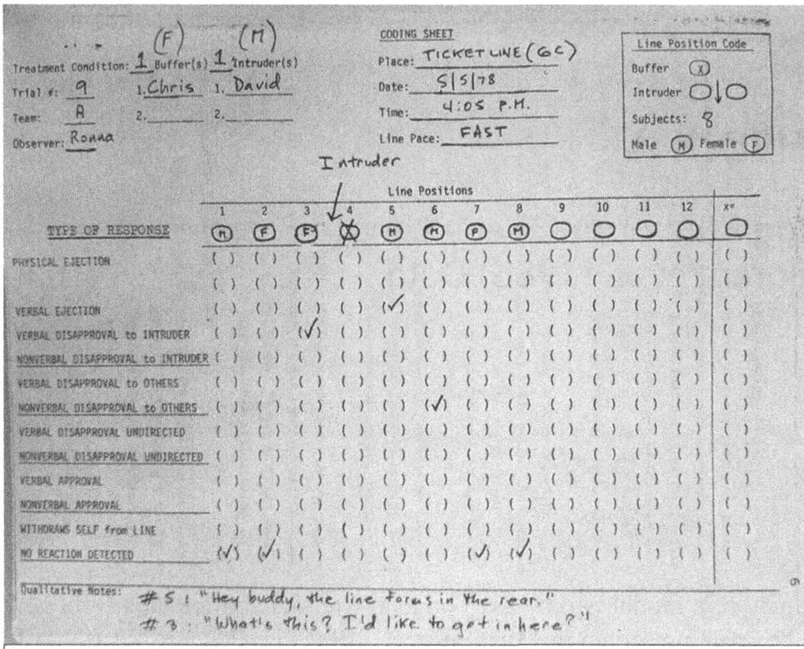

(F) (M)

Treatment Condition: 1 Buffer(s) 1 Intruder(s)
Trial #: 9 1. Chris 1. David
Team: A 2. 2.
Observer: Ronna

CODING SHEET
Place: TICKET LINE (GC)
Date: 5/5/78
Time: 4:05 P.M.
Line Pace: FAST

Line Position Code
Buffer (X)
Intruder O↓O
Subjects: 8
Male (M) Female (F)

Intruder

TYPE OF RESPONSE	1	2	3	4	5	6	7	8	9	10	11	12	X*
	(M)	(F)	(F)	(X)	(M)	(M)	(F)	(M)	( )	( )	( )	( )	( )
PHYSICAL EJECTION	( )	( )	( )	( )	( )	( )	( )	( )	( )	( )	( )	( )	( )
	( )	( )	( )	( )	( )	( )	( )	( )	( )	( )	( )	( )	( )
VERBAL EJECTION	( )	( )	( )	( )	(✓)	( )	( )	( )	( )	( )	( )	( )	( )
VERBAL DISAPPROVAL to INTRUDER	( )	( )	(✓)	( )	( )	( )	( )	( )	( )	( )	( )	( )	( )
NONVERBAL DISAPPROVAL to INTRUDER	( )	( )	( )	( )	( )	( )	( )	( )	( )	( )	( )	( )	( )
VERBAL DISAPPROVAL to OTHERS	( )	( )	( )	( )	( )	( )	( )	( )	( )	( )	( )	( )	( )
NONVERBAL DISAPPROVAL to OTHERS	( )	( )	( )	( )	( )	(✓)	( )	( )	( )	( )	( )	( )	( )
VERBAL DISAPPROVAL UNDIRECTED	( )	( )	( )	( )	( )	( )	( )	( )	( )	( )	( )	( )	( )
NONVERBAL DISAPPROVAL UNDIRECTED	( )	( )	( )	( )	( )	( )	( )	( )	( )	( )	( )	( )	( )
VERBAL APPROVAL	( )	( )	( )	( )	( )	( )	( )	( )	( )	( )	( )	( )	( )
NONVERBAL APPROVAL	( )	( )	( )	( )	( )	( )	( )	( )	( )	( )	( )	( )	( )
WITHDRAWS SELF from LINE	( )	( )	( )	( )	( )	( )	( )	( )	( )	( )	( )	( )	( )
NO REACTION DETECTED	(✓)	(✓)	( )	( )	( )	( )	(✓)	(✓)	( )	( )	( )	( )	( )

Qualitative Notes: #5 : "Hey buddy, the line forms in the rear."
#3 : "What's this? I'd like to get in here?"

Unpublished data obtained from The Stanley Milgram Papers (MS 1406) Manuscripts and Archives, Yale University Library with copy permission from the Estate of Alexandra Milgram.

FIGURE A.1 Original Condition 2 Coding Sheet from the 1978 Intrusion Experiment.

It is important to note that the coding sheet can record the queue position, reaction, and sex of 12 queue members, with overflow information in the last column to the right. After reading Milgram et al. (1986), it might be assumed that the number 12 was selected as the cutoff point so that data could be collected to either support or counter the findings of Harris (1974, p. 564) where all lines used in her experiment were at least 12 persons long. This is not the case though as the following statement indicates,

> *Following completion of our experiment, we learned of a study by Harris (1974) on the frustration-aggression hypothesis (Dollard et al., 1939) that used experimental techniques similar to our own.*
>
> *(Milgram et al., 1986, p. 684, with permission from the American Psychological Association)*

This is important to point out since it implies that queue length selection for any given situation could still be based on random selection. Since Milgram et al. (1986) only provides information to queue position (+4), and the raw data sheets from the Stanley Milgram Papers Manuscripts and Archives, Yale University Library, only provide queue position data out to position (+5), what random selection allows for is another means to systematically obtain approximate minimum queue length data (behind the position of interest) which is required in the case of Condition 4 in Milgram et al. (1986). Given more data exists in the raw data sheets, for consistency, it is the raw data that is used in the calculations contained in this Appendix.

A.1 Basic Queueing Theory and the $M / E_k / 1$ Queue

The focus of this analysis will be on Condition 4, which has three to four members ahead of the intrusion point, and of the 23 trials conducted, 17 of the trials had members out to at least queue position (+5). This indicates that there were likely several queue members consistently beyond queue position (+5), which are not accounted for.

It was shown using data from Schmitt et al. (1992) that the number of queue members behind the position of interest is critical to modeling probability of reaction by the member at the position of interest. An analytic means is therefore needed to systematically approximate the number of queue members behind the position of interest, who are otherwise not accounted for. Based on the approach developed and documented in this appendix, only Condition 4 warrants modification. Condition 4 went from a $Q(2|0|4)$ to $Q(2|0|11.3)$, and Condition 6 will remain the same since it otherwise only would have gone from $Q(2|2|2)$ to a $Q(2|2|2.1)$. All remaining Conditions from Milgram et al. (1986) remain unchanged on the basis of available data regarding number of members behind the position of interest and their probability. Therefore, Condition 4 being the most significant will be used to demonstrate the technique and to document the approach. All other conditions used the same approach but differed only to accommodate number of buffers and unique data pertaining to each of their trial results for the given Condition.

To begin, envision yourself joining a line (a queue) in a grocery store at the end of a five-member queue, including the member currently being served. There is a server at the cash register, which processes the groceries each member is buying, sums up the total cost, bags the groceries, and then takes payment from the queue member being served. Once the queue member being served has paid and all the groceries bagged, then that member leaves the queue. The queue member who was immediately behind him or her is now served next with the process beginning all over again.

After waiting in line for 1 minute, another person joins the queue at the end: three minutes later another, and then 2 minutes later a third so that there are now three queue members behind you when you reach the server at the cash register. In this situation, the sample mean arrival rate (λ) is calculated from three arrivals divided by six minutes, or $\lambda = 0.5$ arrivals per minute. Since your arrival, the five members originally ahead of you were served over the six-minute period, so the sample mean service rate (μ) is five members served divided by the six minutes required to process them all. This results in a single server sample mean rate of $\mu = 0.83$ queue members served per minute. If the arrival rate had been greater than the service rate, then the line would grow infinitely long over an infinite amount of time, or people would just end up leaving the line for a faster one – the latter being more likely.

The Poisson distribution often represents or approximates the specific occurrence of arrival times in many natural examples. In heuristic terms, the postulates that point toward justifiable use of the Poisson distribution for arrival rate in the queue are (Bhattacharyya and Johnson, 1977, p. 157):

1. <u>Independence</u>: The number of events that occur in one period does not affect the number of events that occur in another period (i.e., arrivals are independent).

2. <u>Lack of clustering</u>: Two or more events (arrivals) cannot occur at the same time (i.e., two people cannot occupy the same space at the same time).
3. <u>Rate</u>: The average number of events (arrivals) per unit time is a constant, denoted by λ, and λ does not change with time assuming a homogenous situation (i.e., same conditions and social situation).

What makes the Poisson distribution even more attractive mathematically is that the distribution of time between arrivals is exponentially distributed. This represents a simple embedded Markov process which makes model development much easier. It is not necessary to provide equations since the Poisson distribution is well known, and discussing the attributes of the Markov process is not necessary, other than to indicate it allows use of differential-difference equations, the type we will be demonstrating the development of shortly.

Typically, a person acting as a server at the ticket-counter or grocery store takes an average amount of time to process a queue member – with variation based on the number of items being purchased and the preparedness of the queue member being served. Conversely, if the server is a machine such as Ticketron or a modern ATM, then there will be a mean service time with some variation depending on the person using the machine. So, though service times exist that are exponential, most will be more symmetrical in nature. The Erlang Type k distribution provides this necessary versatility in representation and is still as mathematically tractable as the exponential distribution.

As the integer value for k increases, the Erlang distribution becomes more symmetrical. As $k \to \infty$, the Erlang becomes deterministic with mean value $\frac{1}{\mu}$ (Gross and Harris, 1985, pp. 171–172). Since the sum of k independently and identically distributed exponential random variables, all with mean $\frac{1}{k \cdot \mu}$, yields an Erlang Type k distribution, the Erlang Type k retains some Markovian properties that can be taken advantage of.

It is not the intention to turn this appendix into a minor course on queueing theory, but to provide the steady-state difference equations necessary to calculate the probability p_n of n queue members in an $M / E_k / 1$ queue at any moment or, in this case, the probability of n queue members behind the position of interest. Gross and Harris (1985, p. 175) provide the steady-state difference equations, leaving their solution to the reader. This appendix will spare the reader that effort. Define $p_{n,i}$ as the probability of n queue members being behind the position of interest at any one time, where each member is transitioning from, transitioning to, or remaining in the ith phase of a k-phase service process.

$$0 = -\lambda \cdot p_0 + k \cdot \mu \cdot p_{1,1} \quad \textit{solves for } p_{1,1}\left(p_0\right)$$

$$0 = -\left(\lambda + k\mu\right) \cdot p_{1,i} + k \cdot \mu \cdot p_{1,i+1} \quad \textit{solves for } p_{1,i+1}\left(p_{1,i}\right)\left(1 \le i \le k-1\right)$$

$$0 = -\left(\lambda + k\mu\right) \cdot p_{1,k} + k \cdot \mu \cdot p_{2,1} + \lambda \cdot p_0 \quad \textit{solves for } p_{2,1}\left(p_{1,k}, p_0\right)$$

then, for $n \geq 2$ and $1 \leq i \leq k-1$,

$$0 = -(\lambda + k\mu) \cdot p_{n,i} + k \cdot \mu \cdot p_{n,i+1} + \lambda \cdot p_{n-1,i}$$

$$\text{solves for } p_{n,i+1}\left(p_{n,i}, p_{n-1,i}\right)$$

$$0 = -(\lambda + k\mu) \cdot p_{n,k} + k \cdot \mu \cdot p_{n+1,1} + \lambda \cdot p_{n-1,k}$$

$$\text{solves for } p_{n+1,1}\left(p_{n,k}, p_{n-1,k}\right)$$

It is known that the probability of the queue being empty at any moment is $p_0 = 1 - \frac{\lambda}{\mu}$. Using an Erlang Type 2 as an example, the steady-state difference equations yield:

$$p_{1,1} = \frac{\lambda}{2 \cdot \mu} \cdot p_0, \text{ and } p_{1,2} = \frac{\lambda + 2 \cdot \mu}{2 \cdot \mu} \cdot p_{1,1},$$

$$so\ p_1 = p_{1,1} + p_{1,2};$$

$$p_{2,1} = \frac{(\lambda + 2 \cdot \mu) \cdot p_{1,2} - \lambda \cdot p_0}{2 \cdot \mu}, \text{ and } p_{2,2}$$

$$= \frac{(\lambda + 2 \cdot \mu) \cdot p_{2,1} - \lambda \cdot p_{1,1}}{2 \cdot \mu}, so\ p_2 = p_{2,1} + p_{2,2};$$

$$p_{3,1} = \frac{(\lambda + 2 \cdot \mu) \cdot p_{2,2} - \lambda \cdot p_{1,2}}{2 \cdot \mu}, \text{ and } p_{3,2}$$

$$= \frac{(\lambda + 2 \cdot \mu) \cdot p_{3,1} - \lambda \cdot p_{2,1}}{2 \cdot \mu}, so\ p_3 = p_{3,1} + p_{3,2};$$

The pattern continues, which fortunately we only need to calculate out to p_3 at most for the Conditions with no buffers and since the member of interest does not need to be addressed – he or she is always present. As an example, for Condition 4 (no buffers), p_0 represents the probability of no one behind the position of interest, p_1 is the probability of one person behind the position of interest, and so on. Then $1 - (p_0 + p_1 + p_2 + p_3)$ is the probability of four or more members behind the position of interest that should in a perfect world equal $Q(2|0|4) = \frac{17}{23}$ if the arrival rate and service rate were known, $Q(2|0|4)$ was accurate, and the queue system was actually an $M/E_2/1$. But there lies the main problem: we do not know the arrival rate or the service rate, or even what Erlang Type k distribution best fits the service distribution. What we do know, going back to Condition 4 as an example, is $Q(2|0|0), Q(2|0|1), Q(2|0|2), Q(2|0|3),$ and $Q(2|0|4),$ and that is enough.

It was given $p_0 = 1 - \frac{\lambda}{\mu}$. We really do not need to know both λ and μ but only their ratio $\rho = \frac{\lambda}{\mu}$. With this, we want $\left[Q(2|0|0) - p_o(\rho) \right]^2$ to be as small as possible by varying ρ.

Likewise,

$$Minimize \left[Q(2|0|1) - p_1(\rho) \right]^2 \text{ as a function of } \rho, \text{ where}$$

$$p_1(\rho) = \frac{\lambda}{2 \cdot \mu} \cdot p_0 + \frac{\lambda + 2 \cdot \mu}{2 \cdot \mu} \cdot p_{1,1}$$

$$= \frac{\rho}{2} \cdot p_0(\rho) + \frac{\rho + 2}{2} \cdot p_{1,1}(p_0[\rho]).$$

Using this, we just continue

$$\left[Q(2|0|2) - p_2(\rho) \right]^2, \left[Q(2|0|3) - p_3(\rho) \right]^2,$$

$$and \left[Q(2|0|4) - \left(1 - \sum_{n=0}^{3} p_n(\rho) \right) \right]^2.$$

Once we have a value for ρ that provides the best fit to the empirical data under the assumption $k = 2$, then from (Gross and Harris, 1985), we can find the mean number of queue members L behind the position of interest. Where for an Erlang distribution of Type k,

$$L(\rho) = \frac{k+1}{2k} \cdot \frac{\lambda^2}{\mu \cdot (\mu - \lambda)} + \frac{\lambda}{\mu} = \frac{k+1}{2k} \cdot \frac{\rho^2}{(1 - \rho)} + \rho.$$

In the case of Condition 4, the best fit for Erlang Type 2 is $\rho = \frac{94}{100}$ having a least-squares fit of 0.019. This results in an estimated mean number of queue members behind the position of interest of

$$L(0.94) = \frac{2+1}{4} \cdot \frac{0.94^2}{(1 - 0.94)} + 0.94 = 12 \, members.$$

Having derived the steady-state difference equations up to an Erlang Type 10 (it does get tedious and really is not significantly different beyond $k = 10$), the best fit overall occurs using an Erlang Type 10 with $\rho = \frac{95.2}{100}$ having a least-squares fit of 0.007 and a value for mean number of members behind the position of interest $L = 11.3$ such that

$$L\left(0.952\right) = \frac{10+1}{20} \cdot \frac{0.952^2}{\left(1-0.952\right)} + 0.952 = 11.3 \, members.$$

This is an approximation only, but as evidenced by the other Conditions, it tends to be relatively accurate. Based on simulation, the standard deviation is in the region of about 1.5 queue members for 20 observations. So, what the actual value that would have been recorded on the experiment coding sheet may have been a little less or a little more. Making use of the mean value seemed fair, given the liberties already taken to get this far.

As a final note in this appendix, the author visited Grand Central Station in May of 2019. After collecting 30 minutes of data, the representative service time distribution for a single ticket counter was Erlang Type 5. That is not to say that a Type 5 should have been used instead of a Type 10 for Condition 4, but assuming similarity in general process between 1978 and 2019, use of an Erlang distribution would appear appropriate. It may also be argued that the mean queue length was about twice as long for 17 of the 23 Condition 4 queues compared to what was observed in 2019, implying the tellers at the ticket counter probably operated in a more hurried fashion as would the customers, thus reducing variance and making the process somewhat more deterministic (i.e., possibly explaining the Type 5 observed in 2019 instead of the Type 10 which the 1978 date alludes to).

References

Bhattacharyya, R., & Johnson, R. (1977). *Statistical Concepts and Methods*. Wiley Publishing.

Dollard, J., Miller, N.E., Doob, L.W., Mowrer, O.H., & Sears, R.R. (1939). *Frustration and Aggression*. Yale University Press. https://doi.org/10.1037/10022-000

Gross, H., & Harris, C. (1985). *Fundamentals of Queueing Theory* (2nd ed.). John Wiley and Sons.

Harris, M. (1974). Mediators between Frustration and Aggression in a Field Experiment. *Journal of Experimental Social Psychology*, *10*, 561–571. https://doi.org/10.1016/0022-1031(74)90079-1

Milgram, S., Liberty, J., Toledo, R., & Wackenhut, J. (1986). Response to Intrusion into Waiting Lines. *Journal of Personality and Social Psychology*, *51*(4), 683–689. Copyright American Psychological Association. https://doi.org/10.1037/0022-3514.51.4.683

Schmitt, B., Dube, L., & Leclerc, F. (1992). Intrusions Into Waiting Lines: Does the Queue Constitute a Social System? *Journal of Personality and Social Psychology*, *63*(5), 806–815. https://doi.org/10.1037/0022-3514.63.5.806

GLOSSARY OF VARIABLES AND NOTATION

$e \in [0, d]$	Extent social deviation of social event
E	Independent random variable for extent social deviation such that $e \in R_E$
$es \in [0, \infty)$	Unit stimulus sensation magnitude under exponential encoding
ES	Independent random variable for unit sensation magnitude such that $es \in R_{ES}$
$\overline{es}$	Sample mean value of the independent unit stimulus sensation magnitude random variable ES under exponential encoding
es_p	Sensation magnitude value for p stimuli "units" having unit stimuli sensation magnitude es in social noise N_0 under exponential encoding
$es_{p,m}$	Sensation magnitude value for p stimuli "units" having unit stimuli sensation magnitude es in social noise N_0, and m queue members behind the member position of interest under exponential encoding
$f_S(s)$	Probability density function of the random variable S under uniform encoding
$F_S(s_{p,m})$	Cumulative distribution function (CDF) representing the probability of reaction based on sensation magnitude value $s_{p,m}$ under uniform encoding
$G_{ES}(es_{p,m})$	Cumulative distribution function (CDF) representing probability of reaction based on sensation magnitude value $es_{p,m}$ under exponential encoding
$H_{|UI|+1}(UI)$	Proportion reaction probability is reduced based on observed uncertainty information UI
I_0	Absolute stimulus intensity
I or I_n	Stimulus intensity $\geq$ absolute stimulus intensity, with n > 0
ΔI_n	Change in stimulus intensity $\Delta I_n = I_n - I_{n-1}$
$k \notin P$	Indicates a variable k is not an element of the set P
K	Group member social space

N_0	Background social noise intensity random variable
N_e	Minimum social noise intensity for unit stimulus intensity p such that $N_e = \dfrac{p}{e-1}$
N_n	Social noise intensity above background with $N_n = I_n + N_0$
$\widehat{N}_0$	The geometric mean of the background social noise intensity random variable N_0
N_{UA}	Social uncertainty of attraction intensity random variable (i.e., cohesion noise)
$\widehat{N}_{UA}$	The geometric mean of an uncertainty of attraction random variable
p	Real valued general stimulus intensity, typically represented in this work as positive integer units of stimuli (e.g., number of people in a crowd looking up)
PSE	Point of subjective equality
$r \in [0,1]$	Relevance value of social event
R	Independent random variable for relevance $r \in R_R$
R_X	Range of the random variable X such that $x \in R_X = \{x \,\vert\, f_X(x) > 0\}$.
$\mathbb{R}$	All real numbers
$\mathbb{R}^+$	All real numbers greater than zero
$s \in (0,d]$	Unit stimulus sensation magnitude value under uniform encoding of group member social dissonance with zero social noise intensity N_e such that $s \in \{s = r \cdot e \mid r \in [0,1] \, and \, e \in [0,d]\} - \{0\}$
S	Independent random variable for unit stimulus sensation magnitude $s \in R_S$
$\bar{s}$	Sample mean value of the unit stimulus sensation magnitude random variable S
s_p	Sensation magnitude value for p stimuli "units" having unit stimuli sensation magnitude S in social noise N_0 under uniform encoding
$s_{p,m}$	Sensation magnitude value for p stimuli "units" having unit stimuli sensation magnitude S in social noise N_0 and m queue members behind the member position of interest under uniform encoding
Δs_n	Unit change in sensation magnitude $s_n - s_{n-1}$
$\Delta s_{n,m}$	General change in sensation magnitude $s_n - s_m$
UR	Uncertainty reduction (aka, uncertainty-based information), Chapter 3
$y(\vert UI \vert)$	$y(\vert UI \vert) = H_{\vert UI \vert + 1}(UI)$ when all elements of uncertainty information UI are observed to react
$\mathbb{Z}$	All integers
$\mathbb{Z}^0$	All nonnegative integers
$\mathbb{Z}^+$	All positive integers
$z(\vert UI \vert)$	$z(\vert UI \vert) = H_{\vert UI \vert + 1}(UI)$ when all elements of uncertainty information UI are observed to be nonreactive

INDEX

Note: Page numbers in *italics* indicate figures; page numbers in **bold** indicate tables.

Printed in the United States
by Baker & Taylor Publisher Services